Everyday Electronics Data Book

Michael Tooley BA

PC Publishing

PC Publishing
4 Brook Street
Tonbridge
Kent TN9 2PJ

in association with

Wimborne Publishing Limited
6 Church Street
Wimborne
Dorset BH21 1JH

First published 1990

ISBN 1 870775 02 3

British Library Cataloguing in Publication Data

Tooley, Mike
Everyday electronics data book.
1. Electronics
I. Title
621.381

ISBN 1-870775-02-3

Phototypesetting by Scribe Design, Gillingham, Kent
Printed and bound by BPCC Wheatons Ltd, Exeter

Preface

Welcome to the *Everyday Electronics Data Book*! This book explains the concepts, principles and techniques which have everyday relevance in the world of electronics. The information is presented in a succinct and easy to understand format. The book is not a treatise on electronics theory; it is a text which deals with putting principles into practice and represents a fund of practical knowledge which has been accumulated over more than thirty years.

The book has been written for practising (and aspiring) electronic technicians and engineers involved with the design, manufacture, testing and maintenance of electronic equipment. It will undoubtedly also have a broad appeal to specialists in other disciplines (such as avionics and information technology) who need to be aware of basic electronic principles and practice. The book assumes very little previous knowledge and should also meet the needs of the hobbyist and student. In short, anyone involved with the application of electronics will find this book useful.

The book has been organised into sections and sub-sections, each dealing with a major topic. Explanatory text has been kept necessarily brief and to the point. The text and tabulated data is generously interspersed with examples and hints and tips. These have been included to put the text into context and also to help the reader avoid some of the problems and pitfalls that would otherwise be 'put down to experience'.

Readers should not be content to merely read this book and return it to the bookshelf. It should be treated in much the same way as a good recipe book – kept open on the workbench ready for when it is needed!

Michael Tooley

Foreword

The name Michael Tooley will be familiar to many students of electronics. Michael is a highly respected author in the field and continues to produce regular material for *Everyday Electronics* in addition to the many books he has written.

In assembling the data contained in these pages Mike has drawn on his wide experience in teaching electronics at many levels. His expertise is amply demonstrated by the range, depth, clarity and presentation of the information. I believe this book will become a 'standard' text for everyone involved in electronics.

Mike Kenward
Editor
Everyday Electronics

Contents

1 Fundamentals

SI units

The International System of Units (SI) is based upon the following *fundamental units:*

Quantity	*Unit*	*Abbreviation*
current	ampere	A
length	metre	m
luminous intensity	candela	cd
mass	kilogramme	kg
temperature	degree Kelvin	°K
time	second	s
matter	mol	mol

All other units are derived from the fundamental SI units. Many of these *derived units* have their own names and those commonly encountered in electronics are summarised in the table below together with the quantities to which they relate:

Quantity	*Derived unit*	*Abbreviation*	*Equivalent (in terms of fundamental units)*
capacitance	farad	F	$A\ s\ V^{-1}$
charge	coulomb	C	A s
energy	joule	J	N m

continued overleaf

Quantity	*Derived unit*	*Abbreviation*	*Equivalent (in terms of fundamental units)*
force	newton	N	kg m s^{-1}
frequency	hertz	Hz	s^{-1}
illuminance	lux	lx	lm m^{-2}
inductance	henry	H	V s A^{-1}
luminous flux	lumen	lm	cd sr
magnetic flux	weber	Wb	V s
magnetix flux density	tesla	T	Wb m^{-2}
potential	volt	V	W A^{-1}
power	watt	W	J s^{-1}
resistance	ohm	Ω	V A^{-1}

Angular measure

In electronics, angles are measured in either degrees or radians (both units are strictly dimensionless). The radian is defined as the angle subtended at the centre of a circle by an arc having identical length to the radius of the circle. The unit of solid angle is the steradian (i.e. the solid angle subtended at the centre of a sphere by a square having sides identical in length to the radius of the sphere.

A complete circular revolution is equivalent to a rotation of 360 degrees (360°) or 2π radians (note that π is approximately equal to 3.142). Thus one radian is equivalent to $360/2\pi$ degrees or 57.3°. It is often necessary to convert angles expressed in degrees to radians and vice versa. The following rules should assist readers:

Rule 1. To convert from radians to degrees, multiple by 57.3
Rule 2. To convert from degrees to radians, divide by 57.3

Example 1.1
Two a.c. voltages differ in phase by an angle of 0.8 radians. Express this phase difference in degrees.

The phase difference will be 0.8×57.3 or 45.84° (see Rule 1, above).

Example 1.2
A coil rotates through an angle of 90°. Express this in radians.

The angle in radians is found by dividing the number of degrees by 57.3 (see Rule 2, above). Hence the angle in question is 90/57.3 or 1.57 radians.

Example 1.3
An armature rotates at 3000 r.p.m. What is the angular velocity of the armature (expressed in radians per second)?

In one second the armature will revolve 3000/60 or 50 times. Since one revolution is equivalent to 2π radians, the total angle through which the armature turns will be $50 \times 2\pi$ or 100π radians (note that it is quite permissible to express the number of radians in terms of a multiple of π as this will produce an exact answer). The angular velocity is thus 100π radians per second (or approximately 314 radians per second).

Electrical units and symbols

The following units and symbols are commonly encountered in electronics:

Unit	*Abbrev.*	*Symbol*	*Notes*
ampere	A	I	unit of electric current (a current of 1A flows in a conductor when a charge of 1C is transported in a time interval of 1s)
coulomb	C	Q	unit of electric charge or quantity of electricity
farad	F	C	unit of capacitance (a capacitor has a capacitance of 1F when a charge of 1C results in a potential difference of 1V across its plates)
henry	H	L	unit of inductance (an inductor has an inductance of 1H when an applied current changing uniformly at a rate of 1A/s produces a potential difference of 1V across its terminals)

continued overleaf

Unit	*Abbrev.*	*Symbol*	*Notes*
hertz	Hz	f	unit of frequency (a signal has a frequency of 1Hz if one complete cycle occurs in a time interval of 1s)
joule	J	E	unit of energy
ohm	Ω	R	unit of resistance
second	s	t	unit of time
siemen	S	G	unit of conductance (reciprocal of resistance)
tesla	T	B	unit of magnetic flux density (a flux density of 1T is produced when a flux of 1Wb is present over an area of 1 square metre)
volt	V	V	unit of electrical potential (e.m.f. or p.d.)
watt	W	P	unit of power (equal to 1J of energy consumed in a time of 1s)
weber	Wb	Φ	unit of magnetic flux

Multiples and sub-multiples

Many of the fundamental units are somewhat cumbersome for everyday use. Hence the following multiples and sub-multiples are employed:

Prefix.	*Abbrev.*	*Multiplier*
tera	T	10^{12} (=1000000000000)
giga	G	10^{9} (=1000000000)
mega	M	10^{6} (=1000000)
kilo	k	10^{3} (=1000)
(none)	(none)	10^{0} (=1)
centi	c	10^{-2} (=0.01)
milli	m	10^{-3} (=0.001)
micro	μ	10^{-6} (=0.000001)
nano	n	10^{-9} (=0.000000001)
pico	p	10^{-12} (=0.000000000001)

Example 1.4
A current of 0.025A flows in a circuit. Express this in mA.

We can express the current in mA (rather than in A) by simply moving the decimal point three places to the right. Hence 0.025A is the same as 25mA.

Example 1.5
A signal has a frequency of 795kHz. Express this in MHz.

To express the frequency in MHz rather than kHz we need to move the decimal point three places to the left. Hence 795kHz is equivalent to 0.795MHz.

Example 1.6
A capacitor has a value of 27000pF. Express this in μF.

To express the value in μF rather than pF we need to move the decimal point six places to the left. Hence 27000pF is equivalent to 0.027μF (note that we have had to introduce an extra zero before the 2 and after the decimal point).

Conductors and insulators

In metals, electric current is simply the organised movement of free electrons. Materials which readily support the flow of an electric current are known as conductors, whilst those which do not readily permit current flow are known as insulators. Examples of conductors are metals (such as copper, aluminium, gold and silver) whilst examples of insulators are plastics and ceramic materials. The properties of commonly used conductors and insulators are summarised in the tables which follow:

(a) Properties of common metallic conductors

Material	*Resistivity (Ω m)*	*Temperature coefficient (at 20°C)*	*Thermal conductivity (at 20°C)*	*Melting point (°C)*
aluminium	2.7×10^{-8}	4.0×10^{-3}	0.48	660
brass	7.2×10^{-8}	2.0×10^{-3}	0.26	920
Constantan	4.9×10^{-7}	1.0×10^{-5}	0.054	1210

continued overleaf

Material	*Resistivity (Ω m)*	*Temperature coefficient (at 20°C)*	*Thermal conductivity (at 20°C)*	*Melting point (°C)*
copper	1.6×10^{-8}	4.3×10^{-3}	0.918	1083
gold	2.3×10^{-8}	3.4×10^{-3}	0.705	1063
iron (cast)	9.1×10^{-8}	6.0×10^{-3}	0.18	1535
lead	2.0×10^{-7}	4.2×10^{-3}	0.083	327
Nichrome	1.0×10^{-6}	1.7×10^{-4}	0.035	1350
nickel	1.0×10^{-7}	4.7×10^{-3}	0.142	1452
silver	1.5×10^{-8}	4.0×10^{-3}	1.006	960.5
tin	1.3×10^{-7}	4.2×10^{-3}	0.155	231.9
tungsten	5.4×10^{-8}	4.5×10^{-3}	0.476	3370

(b) Properties of common insulators

Material	*Volume resistivity (Ω m)*	*Dielectric constant (100Hz–100MHz)*	*Dielectric strength (kV/mm)*	*Maximum rec. temperature (°C)*
Bakelite	10^{10}	4.4–5.4	11.8	+100
glass (Pyrex)	10^{12}	4.8	13.2	+600
polyester film	10^{13}	2.8–3.7	27.6	+105
polyethylene	10^{14}	2.2	23.0	+60
polypropylene	10^{14}	2.0	23.6	+100
porcelain	10^{13}	5.1–5.9	11.8	+1000
PTFE	$>10^{13}$	2.2	23.6	+250
PVC	10^{8}	3.6–4.0	31.5	+85
Teflon	$>2\times10^{16}$	2.1	110	+200

Current, voltage and resistance

The ability of an energy source (e.g. a battery) to produce a current within a conductor is a measure of its electromotive force (e.m.f.). Whenever an e.m.f. is applied to a circuit a potential difference (p.d.) exists. Both e.m.f. and p.d. are measured in volts (V). In many practical circuits there is only one e.m.f. present (the battery or supply) however a p.d. will be developed across each component in the circuit.

For any conductor, the current flowing is directly proportional to the e.m.f. applied. The current flowing will also be dependent on the physical dimensions (length and cross-sectional area) and material of which the conductor is composed. The amount of current that will flow in a conductor when a given e.m.f. is applied is inversely proportional to its resistance. Resistance, therefore, may be thought of as an opposition to current flow; the higher the resistance the lower the current that will flow (assuming that the applied e.m.f. remains constant).

Ohm's law

Provided that temperature does not vary, the ratio of p.d. across the ends of a conductor to the current flowing in the conductor is a constant. This relationship is known as *Ohm's law* and this leads us to the conclusion that:

$V/I = \text{a constant} = R$

where V is the p.d. in volts (V), I is the current in amperes (A), and R is the resistance in ohms (Ω) (see Figure 1.1).

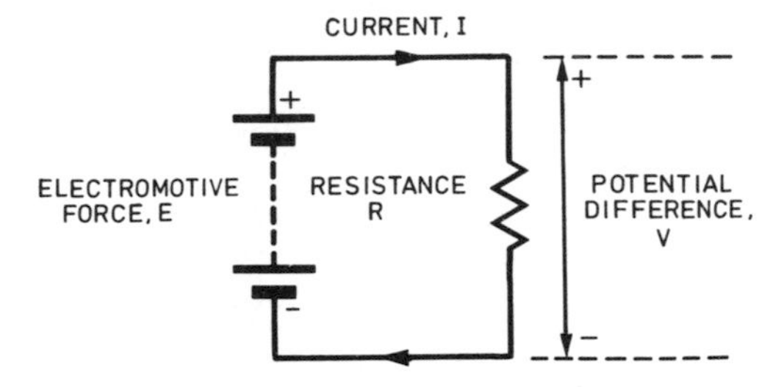

Figure 1.1 E.m.f., current and potential difference

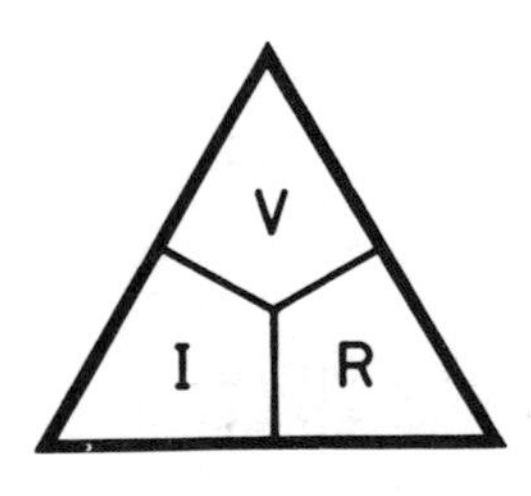

Figure 1.2 Relationship between V, I and R

The formula may be arranged to make V, I or R the subject, as follows:

$V = I \times R \quad I = V/R \quad \text{and} \quad R = V/I$

The triangle shown in Figure 1.2 should help readers remember these three important relationships.

Example 1.7
A 12Ω resistor is connected to a 6V battery. What current will flow in the resistor?

Here we must use $I = V/R$ (where V = 6V and R = 12Ω):

$I = V/R = 6V/12\Omega = 0.5A$ (or 500mA)

Hence a current of 500mA will flow in the resistor.

Example 1.8
A current of 100mA flows in a 56Ω resistor. What voltage drop (p.d.) will be developed across the resistor?

Here we must use $V = I \times R$ and ensure that we work in units of volts (V), amps (A), and ohms (Ω).

$V = I \times R = 0.1A \times 56\Omega = 5.6V$

(Note that 100mA is the same as 0.1A)

Hence a p.d. of 5.6V will be developed across the resistor.

Example 1.9
A voltage drop of 15V appears across a resistor in which a current of 1mA flows. What is the value of the resistance?

$R = V/I = 15V/0.001A = 15000\Omega = 15k\Omega$

Note that it is often more convenient to work in units of mA and V which will produce an answer directly in kΩ, i.e.

$R = V/I = 15V/1mA = 15k\Omega$

Resistance and resistivity

The resistance of a metallic conductor is directly proportional to its length and inversely proportional to its area. The resistance is also directly proportional to its *resistivity* (or *specific resistance*). Resistivity is defined as the resistance measured between the opposite faces of a cube having sides of 1cm.

The resistance, R, of a conductor is thus given by the formula:

$$R = \rho \times l/A$$

where R is the resistance (in Ω), ρ is the resistivity (in Ω m), l is the length (in m), and A is the area in m^2.

Example 1.10

An inductor consists of a 8m length of copper wire having a cross sectional area of $1mm^2$. Determine the resistance of the inductor.

We will use the formula, $R = \rho\ l/A$

The value of ρ for copper is 1.6×10^{-8} Ωm given earlier in the table which shows the properties of common metallic conductors. The length of the wire is 4m whilst the area is $1mm^2$ or $1 \times 10^{-6}\ m^2$ (note that we must be consistent in using units of metres for length and square metres for area). Hence the resistance of the inductor will be given by:

$$R = 1.6 \times 10^{-8} \times \frac{8}{1 \times 10^{-6}}$$

Thus $R = 12.8 \times 10^{-2}$ or 0.128 Ω.

Example 1.11

A copper cable of length 20m and cross sectional area $1mm^2$ carries a current of 5A. Determine the voltage drop between the ends of the cable.

First we must find the resistance of the cable (as in Example 1.12):

$$R = \rho\ l/A = 1.6 \times 10^{-8} \times \frac{20}{1 \times 10^{-6}} = 0.32\Omega.$$

The voltage drop can now be calculated using Ohm's law:

$$V = I \times R = 5 \times 0.32 = 1.6V$$

Hence 1.6V will be dropped between the ends of the cable.

Energy and power

Energy is the ability to do work, and power is the rate at which work is done. Electrical energy may be stored in components such as capacitors and inductors or converted into various other forms of energy by components such as resistors (heat), piezoelectric transducers (sound), LEDs (light).

The unit of energy is the *joule* (J). In electronic circuits, power is measured in *watts* (W), and a power of 1W results from energy being used at the rate of 1J per second. The power in a circuit is equivalent to the product of voltage and current. Hence:

$$P = I \times V$$

where P is the power in watts (W), I is the current in amperes (A), and V is the voltage in volts (V).

The formula may be arranged to make P, I or V the subject, as follows:

$$P = I \times V \quad I = P/V \quad \text{and} \quad V = P/I$$

The triangle shown in Figure 1.3 shows these relationships.

The relationship, $P = I \times V$, may be combined with that which results from Ohm's law ($V = I \times R$), to produce several further relationships which are summarised in Figure 1.4.

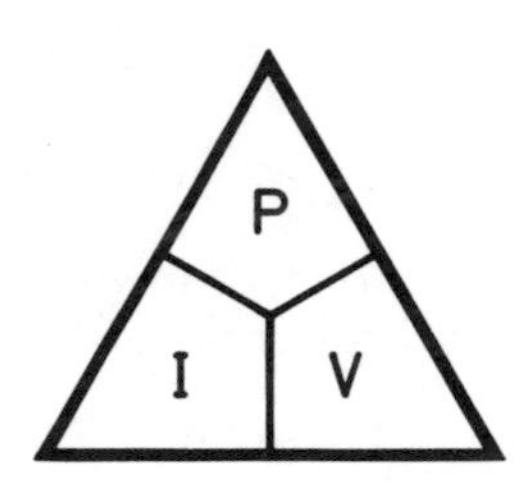

Figure 1.3 Relationship between P, I and V

V	I	R	P
IR	$\frac{V}{R}$	$\frac{V}{I}$	$\frac{V^2}{R}$
$\frac{P}{I}$	$\frac{P}{V}$	$\frac{V^2}{P}$	$I^2 R$
$\sqrt{PR}$	$\sqrt{\frac{P}{R}}$	$\frac{P}{I^2}$	IV
VOLTS	AMPS	OHMS	WATTS

Figure 1.4 Various useful relationships summarised

Example 1.12
A current of 1.5A is drawn from a 3V battery. What power is supplied?

Here we must use $P = I \times V$ (where $I = 1.5A$ and $V = 3V$):

$P = I \times V = 1.5A \times 3V = 4.5W$

Hence a power of 4.5W is supplied.

Example 1.13
A voltage drop of 4V appears across a resistor of 100Ω. What power is dissipated in the resistor?

Here we use $P = V^2/R$ (where $V = 4V$ and $R = 100\Omega$):

$P = V^2/R = (4V \times 4V)/100\Omega = 0.16W$

Hence the resistor dissipates a power of 0.16W (or 160mW).

Example 1.14
A current of 20mA flows in a 1kΩ resistor. What power is dissipated in the resistor?

Here we use $P = I^2 \times R$ but, to make life a little easier, we will work in mA and kΩ (in which case the answer will be in mW):

$P = I^2 \times R = (20mA \times 20mA) \times 1k\Omega = 400mW$

Thus a power of 400mW is dissipated by the resistor.

Direct and alternating current

Direct currents are currents which, even though their magnitude may vary, essentially flow only in one direction. The conventional flow of current in a circuit is from the point of more positive potential to the point of greatest negative potential (note that electron movement is in the opposite direction!). Direct currents result from the application of a direct e.m.f. (derived from batteries or d.c. supply rails). An essential characteristic of such supplies is that the applied e.m.f. does not change its polarity (even though its value may be subject to some fluctuation).

Direct currents are *unidirectional*. Alternating currents, on the other hand, are *bidirectional* and flow first in one direction and then in the other. The polarity of the e.m.f. which produces an alternating current must consequently also be changing.

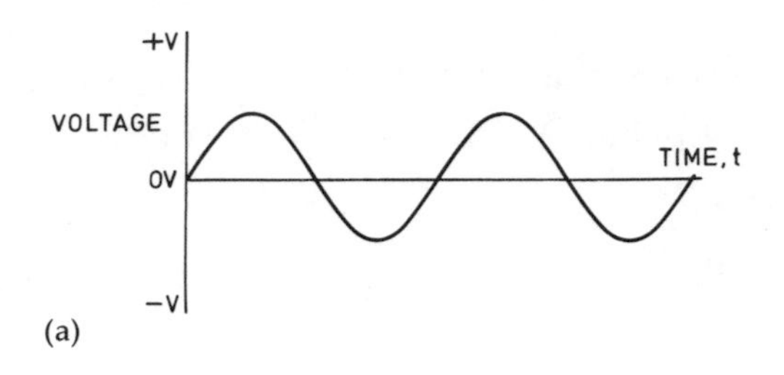

(a)

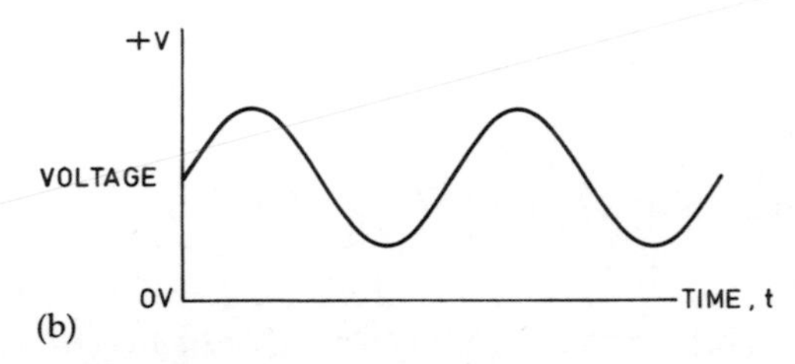

(b)

Figure 1.5 Alternating voltages: (a) bipolar; (b) unipolar

Alternating currents produce alternating potential differences (voltages) in the circuits in which they flow. Furthermore, in many circuits, alternating voltages frequently appear superimposed on direct voltage levels. The resulting voltage may be *unipolar* (i.e. always positive or always negative) or *bipolar* (i.e. partly positive and partly negative). Figure 1.5 illustrates this point.

Waveforms

A graph showing the variation of voltage or current present in a circuit is known as a *waveform*. There are many common types of waveform encountered in electronic circuits including *sine* (or *sinusoidal*), *square, triangle, ramp* or *sawtooth* (which may be either positive or negative going), and *pulse*. The waveforms for speech and music are often referred to as *complex* since they invariably comprise many components at widely differing frequencies. Pulse waveforms are often categorised as either *repetitive* or *non-repetitive* (the former comprises a pattern of pulses which regularly repeats whilst the latter comprises pulses which constitute a unique event). Several of the most common waveform types are shown in Figure 1.6.

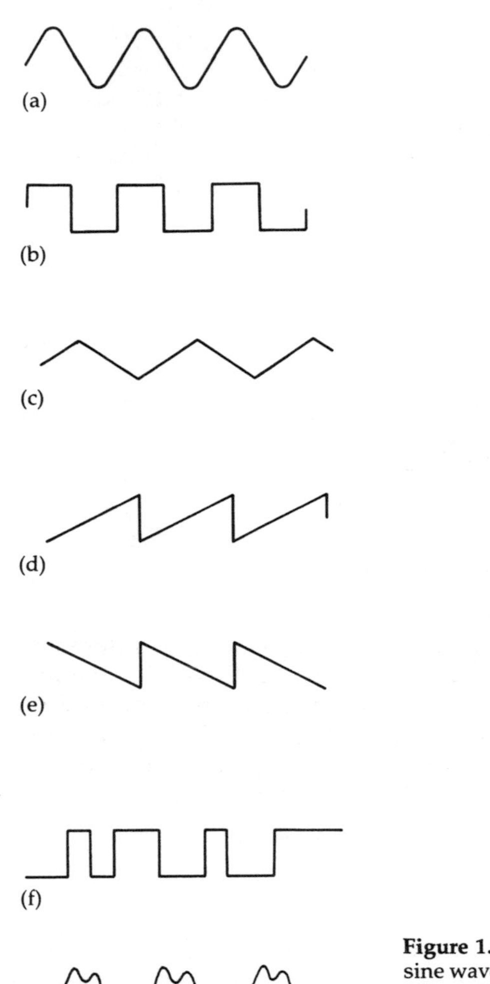

Figure 1.6 Several common waveforms: (a) sine wave; (b) square wave; (c) triangle wave; (d) positive ramp (sawtooth); (e) negative ramp (sawtooth); (f) pulse waveform; (g) complex waveform

Frequency and periodic time

The frequency of a (repetitive) waveform is the number of cycles of the waveform which occur in unit time. Frequency is expressed in *hertz* (Hz). A frequency of 1Hz is equivalent to one cycle per

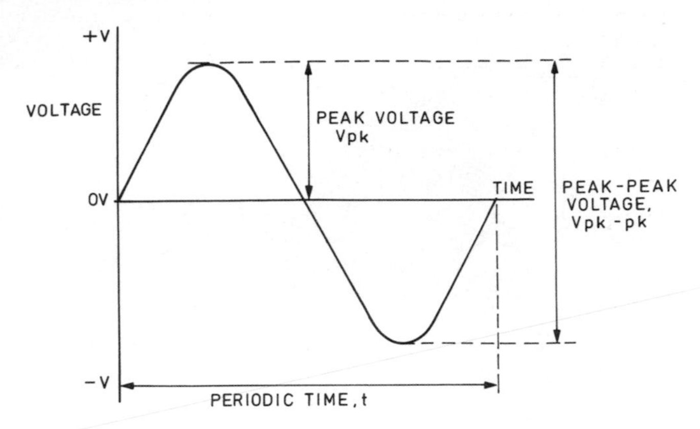

Figure 1.7 Periodic time, peak and peak-peak values

second. Thus if a wave alternates at a rate of 50 cycles per second, its frequency is 50Hz.

The *periodic time* (or period) of a waveform is the time taken for one complete cycle of the wave (see Figure 1.7). The relationship between periodic time and frequency is thus:

$t = 1/f$ or $f = 1/t$

where t is the periodic time (in seconds) and f is the frequency (in Hz). The following table gives some representative values of frequency and periodic time:

Frequency	*Periodic time*
1Hz	1s
2Hz	500ms
5Hz	200ms
10Hz	100ms
20Hz	50ms
50Hz	20ms
100Hz	10ms
200Hz	5ms
500Hz	2ms
1kHz	1ms
2kHz	500μs
5kHz	200μs
10kHz	100μs

Frequency	*Periodic time*
20kHz	50μs
50kHz	20μs
100kHz	10μs
200kHz	5μs
500kHz	2μs
1MHz	1μs

Example 1.5
A waveform has a frequency of 30Hz. What is the periodic time of the waveform?

Using t = 1/f (where f = 30Hz):

t = 1/30Hz = 0.0333s (or 33.3ms).

Hence the waveform has a periodic time of 33.3ms.

Example 1.16
A waveform has a periodic time of 4ms. What is its frequency?

Using f = 1/t (where t = 4ms):

f = 1/4ms = 250Hz

Hence the frequency is 250Hz.

Average, peak, peak-peak, and r.m.s. values

The average value of a waveform which swings symmetrically above and below zero will obviously be zero when measured over a long period of time. Hence *average* values of currents and voltages are invariably taken over one complete *half-cycle* (either positive or negative) rather than over one complete *full-cycle* (which would result in an average value of zero).

The *amplitude* (or *peak value*) of a waveform is a measure of the extent of its voltage or current excursion from the resting value (*usually* zero). The *peak-to-peak* value for a wave which is symmetrical about its resting value is *twice* its peak value (see Figure 1.7).

The *r.m.s.* (or *effective*) value of an alternating voltage or current is the value which would produce the same heat energy in a resistor as a direct voltage or current of the same magnitude. The r.m.s. value of a waveform is very much dependent upon its shape. R.M.S. values are, therefore, only meaningful when dealing with a waveform of known shape. Where the shape of a waveform is not specified, r.m.s. values are normally assumed to refer to sinusoidal conditions.

For a given waveform, a set of fixed relationships exist between average, peak, peak-peak and r.m.s. values. The required multiplying factors are summarised below for sinusoidal voltages and currents:

		Wanted quantity			
		Average	*Peak*	*Peak-peak*	*R.M.S.*
	Average	1	1.57	3.14	1.11
Given	Peak	0.636	1	2	0.707
quantity	Peak-peak	0.318	0.5	1	0.353
	R.M.S.	0.9	1.414	2.828	1

Example 1.17

A sinusoidal voltage has an r.m.s. value of 240V. What is the peak value of the voltage?

The corresponding multiplying factor (found from the table above) is 1.414.

Hence $V_{pk} = 1.414 \times V_{rms}$

Thus $V_{pk} = 1.414 \times 240V_{rms} = 339.4V_{pk}$.

Example 1.18

An alternating current has a peak-peak value of 50mA. What is its r.m.s. value?

The corresponding multiplying factor (found from the table above) is 0.353.

Hence $I_{rms} = 0.353 \times I_{pk\text{-}pk}$

Thus $I_{rms} = 0.353 \times 50mA_{pk\text{-}pk} = 17.65mA_{pk\text{-}pk}$

Example 1.19

A sine wave of $10V_{pk\text{-}pk}$ is applied to a resistor of 1kΩ. What value of r.m.s. current will flow in the resistor?

This problem needs to be solved in two stages. First we will determine the peak-peak current in the resistor and then convert this value into a corresponding r.m.s. quantity.

Since $I = V/R$ then $I_{pk\text{-}pk} = V_{pk\text{-}pk}/R$
Hence $I_{pk\text{-}pk} = 10V_{pk\text{-}pk}/1k\Omega = 10mA_{pk\text{-}pk}$

The required multiplying factor (peak-peak to r.m.s.) is 0.353. Thus:

$I_{rms} = 0.353 \times I_{pk\text{-}pk} = 0.353 \times 10mA_{pk\text{-}pk} = 3.53mA_{rms}$

Hints and tips

★ When performing calculations of currents, voltages and resistances, it is seldom necessary to work with an accuracy of better than 1% simply because component tolerances are invariably somewhat greater than this.

★ In calculations involving Ohm's law, it is often convenient to work in units of kΩ and mA (or MΩ and μA) in which case potential differences will be expressed directly in V.

★ When multiplying two values which are expressed using exponents, it is simply necessary to *add* the exponents. As an example:

$(2 \times 10^2) \times (3 \times 10^6) = (2 \times 3) \times 10^{(2+6)} = 6 \times 10^8$

★ When dividing two values which are expressed using exponents, it is simply necessary to *subtract* the exponents. As an example:

$(4 \times 10^6) \div (2 \times 10^4) = 4/2 \times 10^{(6-4)} = 2 \times 10^2$

★ Always take care to express the units, multiples and submultiples in which you are working. Failure to do this can result in (at best) confusion and (at worst) catastrophic failure of circuits and components!

★ A good scientific calculator is a very worthwhile investment which can greatly simplify the task of performing calculations in electronic circuits. When purchasing such a device, check that it can cope with angles expressed in *either* radians or degrees and that it has square, square root, log, and π function/constant keys.

2 Passive components

Resistors

Resistors provide us with a means of controlling the current and/or voltage present in a circuit. Typical applications involve the provision of bias potentials and currents for transistor amplifiers, converting the collector or emitter output current of a transistor into a corresponding output voltage drop, and providing a predetermined value of attenuation.

The electrical characteristics of a resistor are largely determined by the material of which it is composed and its construction. Factors which must be considered when selecting a resistor for a particular application normally include:

(a) The required *value* of resistance (expressed in Ω, kΩ or MΩ)
(b) The desired *accuracy* or *tolerance* (quoted as the maximum permissible percentage deviation from the marked value).
(c) The *power rating* (which must be equal to, or greater than, the maximum expected power dissipation)
(d) The *temperature coefficient* of the resistance (usually expressed in parts per million, ppm, per unit temperature change).
(e) The *stability* of the resistor. Usually expressed in terms of the long or short-term percentage variation of resistance which occurs under specified physical and electrical operating conditions.
(f) *Noise* performance (usually expressed in terms of the equivalent noise voltage generated by the resistor under a specified set of physical and electrical conditions).

Note that, rather than quote the manufacturers' specifications for items (e) and (f) most suppliers simply state 'high-stability' or 'low-noise'. In addition, resistors offering a tolerance of ±2% (or better) are often referred to as being of 'close-tolerance'. In general, high-stability, low-noise and close-tolerance resistors are only required in more critical applications. Such applications include the initial stages of amplifiers which cater for signals of very small amplitude as well as the input stages of measuring and test equipment. There is, of course, no reason why high performance resistors should not be used for less critical applications other than their significantly higher cost! The following table summarises the properties of commonly available types of resistor:

Resistor type	*Carbon composition*	*Carbon film*	*Metal film*	*Metal oxide*
Resistance range (Ω)	2.2 to 1M	10 to 10M	1 to 1M	10 to 1M
Typical tolerance (%)	±10	±5	±1	±2
Power rating (W)	0.125 to 1	0.25 to 2	0.125 to 0.5	0.25 to 0.5
Temperature coefficient (ppm/°C)	+1200	−250	+50 to +100	+250
Stability	poor	fair	excellent	excellent
Noise performance	poor	fair	excellent	excellent
Ambient temperature range (°C)	−40 to +105	−45 to +125	−55 to +125	−55 to +155
Typical unit cost	2p	2p	3p	4p
Typical applications	power supplies and large-signal amplifiers only	general purpose	small-signal amplifiers, test and measuring equipment	

Resistor type	*Ceramic wirewound*	*Vitreous enamel wirewound*	*Aluminium clad wirewound*
Resistance range (Ω)	0.47 to 22k	0.1 to 22k	0.1 to 1k
Typical tolerance (%)	±5	±5	±5
Power rating (W)	4 to 17	2 to 4	25 to 50 (note 1)

continued overleaf

Resistor type	*Ceramic wirewound*	*Vitreous enamel wirewound*	*Aluminium clad wirewound*
Temperature coefficient (ppm/°C)	±250	±75	±50
Stability	good	good	good
Noise performance	(note 2)	(note 2)	(note 2)
Ambient temperature range (°C)	−55 to +200	−55 to +200	−55 to +200
Typical unit cost	25p	35p	80p
Typical applications	power supplies and large signal amplifiers only	general purpose power dissipation	small-signal amplifiers, test and measuring equipment

Note 1 Requires mounting on a substantial heatsink to dissipate the quoted power level. Derate by approximately 50% for mounting in free air.
Note 2 Not normally an important consideration in applications involving this type of resistor.

Preferred values

The resistance marked on the body of a resistor is only a guide to its actual value of resistance. A resistor marked 100Ω and having a tolerance of ±10% will, for example, have a value which falls within the range 90Ω to 110Ω. If a particular application requires a resistance of, for example 105Ω, a ±10% tolerance resistor of 100Ω will be perfectly adequate. If, however, we require a value of 101Ω, then it would be necessary to obtain a 100Ω resistor having a marked value of 100Ω and a tolerance of ±1%.

Resistors are available in several series of fixed decade values, the number of values provided with each series being governed by the tolerance involved. In order to cover the full range of resistance values using resistors having a ±20% tolerance it will be necessary to provide six basic values (known as the *E6 series*). More values will be required in the series which offers a tolerance of ±10% and consequently the *E12 series* provides twelve basic values. The *E24 series* for resistors of ±5% tolerance provides no less than 24 basic values and, as with the E6 and E12 series, decade multiples (i.e. ×1, ×10, ×100, ×1k, ×10k, ×100k and ×1M) of the basic series.

	Series	
E6	*E12*	E24
1.0	1.0	1.0
		1.1
	1.2	1.2
		1.3
1.5	1.5	1.5
		1.6
	1.8	1.8
		2.0
2.2	2.2	2.2
		2.4
	2.7	2.7
		3.0
3.3	3.3	3.3
		3.6
	3.9	3.9
		4.3
4.7	4.7	4.7
		5.1
	5.6	5.6
		6.2
6.8	6.8	6.8
		7.5
	8.2	8.2
		9.1

Example 2.1
An amplifier stage requires an accurate bias current of 100μA (±10%) derived from a 5V d.c. supply rail. What value and type of resistor should be used in this application?

The value of resistance required must first be calculated using Ohm's law:

R = V/I = 5V/100μA = 50kΩ.

The nearest preferred value from the E24 series is 51kΩ (which will actually produce a current of 98μA i.e. within 2% of the desired value). If a resistor of ±5% tolerance is used, the bias current will be within the range 93μA to 103μA (well within the ±10% accuracy specified). The power dissipated in the resistor (calculated using $P = I \times V$) will be very small (500μW) and thus a 0.25W resistor will be perfectly adequate for this application. A carbon film resistor should be adequate for this application however, if the amplifier operates with small signals and noise performance is critical then a metal oxide resistor would be preferable.

Hints and tips

★ Resistor power ratings are related to operating temperatures and resistors should be derated at high temperatures. Where reliability is important resistors should be operated at well below their normal maximum power dissipation.

★ Where a number of resistors of identical value are required, it can often be more cost-effective to make use of thick film resistor networks rather than discrete components. Thick film networks are available in DIL and SIL packages with either four individual or seven/eight commoned resistors in the E12 series between 33Ω and 100Ω (see Figure 2.1).

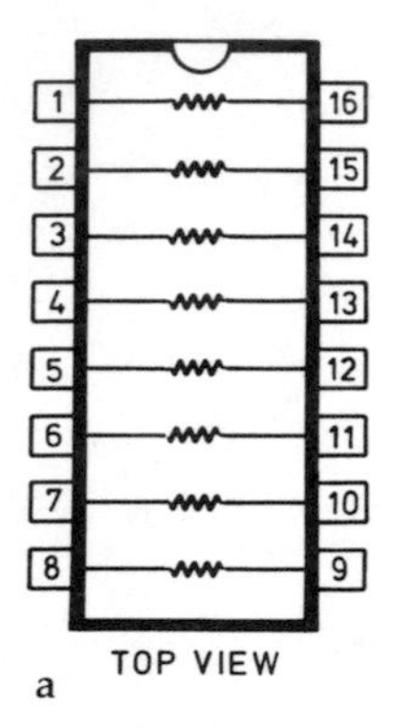

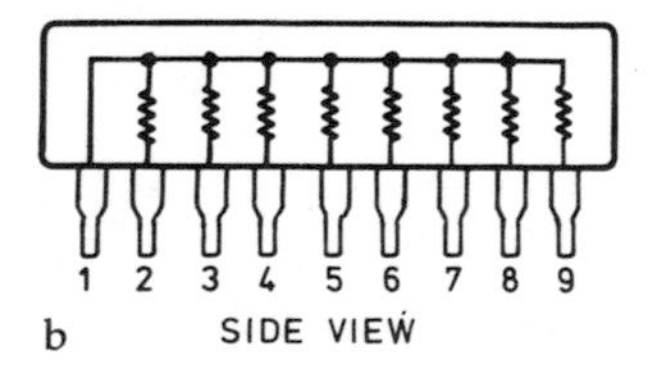

Figure 2.1 Typical DIL and SIL resistor packages. (a) eight individual resistors in a DIL package; (b) eight commoned resistors in a SIL package

★ Wirewound resistors are inductive at high frequencies and this generally precludes their use at frequencies in excess of 100kHz. At much higher frequencies (i.e. greater than 30MHz) carbon and metal film resistors also suffer from stray inductance (attributable to their axial connecting leads). In such applications, the lengths of resistor leads should be kept to an absolute minimum in order to minimise the effects of unwanted stray inductance.
★ The insulating qualities of vitreous enamel coatings tend to become impaired at high temperatures. In applications involving operation at, or near, the maximum permitted dissipation, this type of resistor, should *not* be left in contact with any conducting surface.

Resistor colour codes

Carbon and metal oxide resistors are invariably marked with colour codes which indicate their value and tolerance. Two methods of colour coding are in common use; one involves four coloured bands (see Figure 2.2) whilst the other uses five colour bands (see Figure 2.3).

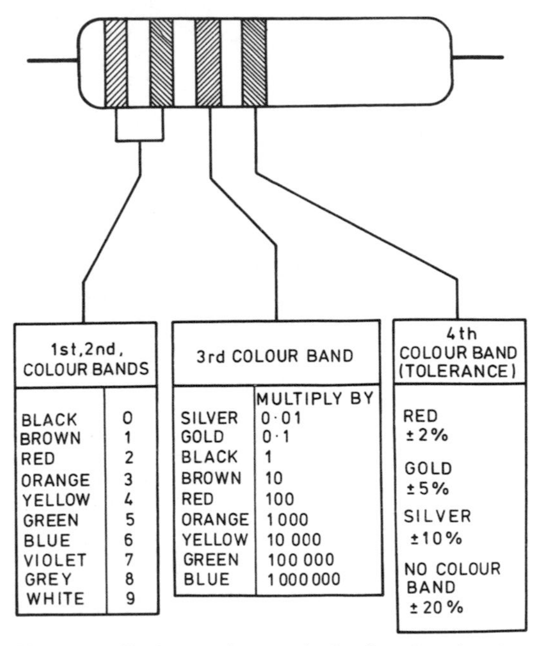

Figure 2.2 Resistor colour code for four-band resistors

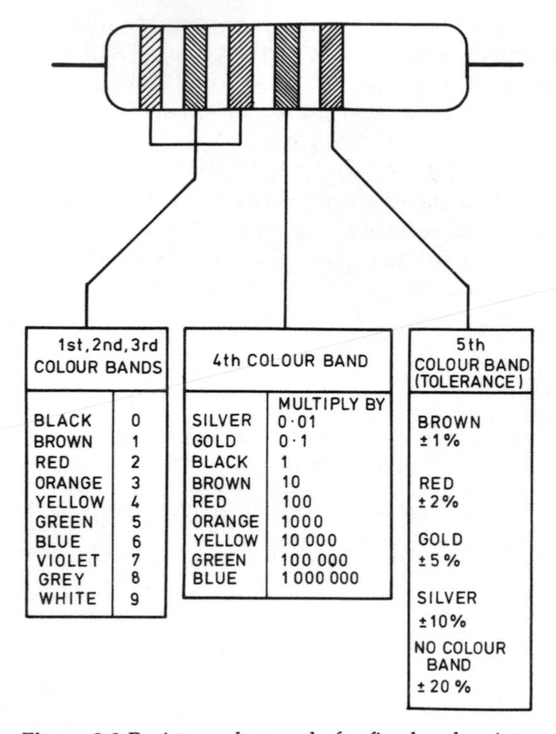

Figure 2.3 Resistor colour code for five-band resistors

Example 2.2

A resistor is marked coded with the following coloured stripes: brown, black, red, gold. What is its value and tolerance?

First digit: brown = 1
Second digit: black = 0
Multiplier: red = 2 (× 100)
Value: 10 × 100 = 1000Ω = 1kΩ
Tolerance: gold = ±5%

Example 2.3

A resistor is marked coded with the following coloured stripes: blue, grey, orange, silver. What is its value and tolerance?

First digit: blue = 6
Second digit: grey = 8

Multiplier: orange = 3 ($\times$ 1000)
Value: 68 $\times$ 1000 = 68000Ω = 68kΩ
Tolerance: silver = ±10%

Example 2.4
A resistor is marked coded with the following coloured stripes: orange, orange, silver, silver. What is its value and tolerance?

First digit: orange = 3
Second digit: orange = 3
Multiplier: silver = ÷ 100
Value: 33/100 = 0.33Ω
Tolerance: silver = ±10%

Example 2.5
A resistor is marked coded with the following coloured stripes: red, yellow, black, black, red. What is its value and tolerance?

First digit: red = 2
Second digit: yellow = 4
Third digit: black = 0
Multiplier: black = 0 ($\times$ 1)
Value: 240 $\times$ 1 = 240Ω
Tolerance: red = ±2%

BS 1852 coding

Some types of resistor have values marked using the system of coding defined in BS 1852. This system involves marking the position of the decimal point with a letter to indicate the multiplier concerned as follows:

Letter	*Multiplier*
R	1
K	1000
M	1000000

A further letter is then appended to indicate the tolerance as shown below:

Letter	*Tolerance*
F	±1%
G	±2%
J	±5%
K	±10%
M	±20%

Example 2.6
A resistor is marked coded with the legend 4R7K. What is its value and tolerance?

4.7Ω ±10%

Example 2.7
A resistor is marked coded with the legend 330RG. What is its value and tolerance?

330Ω ±2%

Example 2.8
A resistor is marked coded with the legend R22M. What is its value and tolerance?

0.22Ω ±20%

Series and parallel combinations of resistors

Fixed resistors may be arranged in a variety of series and parallel combinations in order to obtain particular values of resistance. Some representative arrangements are shown in Figures 2.4 and

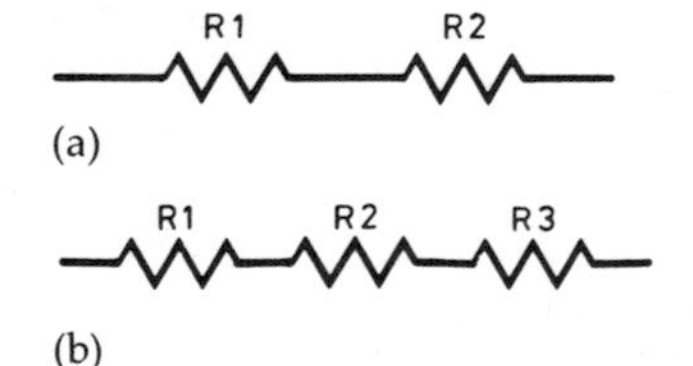

Figure 2.4 Series connected resistors: (a) two resistors in series; (b) three resistors in series

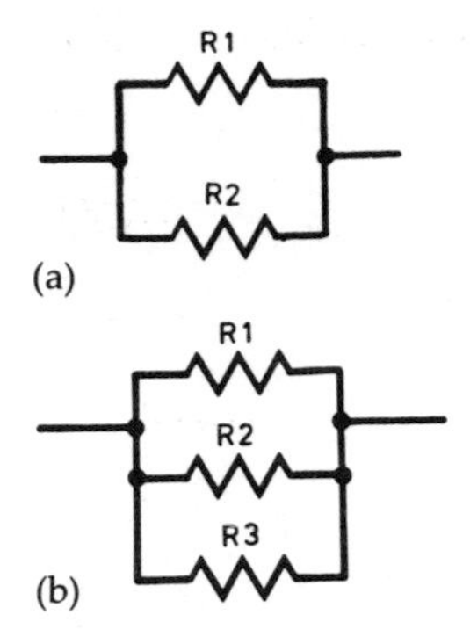

Figure 2.5 Parallel connected resistors: (a) two resistors in parallel; (b) three resistors in parallel

2.5. The effective resistance of the series circuits is simply the sum of the individual resistances. Hence, for figure 2.4(a):

$$R = R1 + R2$$

whilst in Figure 2.4(b):

$$R = R1 + R2 + R3$$

In the case of the parallel connected resistors in Figure 2.5, the reciprocal of the effective resistance of the combination is equal to the sum of the reciprocals of the individual resistances. Hence, in Figure 2.5(a):

$$\frac{1}{R} = \frac{1}{R1} + \frac{1}{R2}$$

whilst in Figure 2.5(b):

$$\frac{1}{R} = \frac{1}{R1} + \frac{1}{R2} + \frac{1}{R3}$$

Where only two resistors are connected in parallel (as in Figure 2.5(a)) the formula can be re-arranged as follows:

$$R = \frac{R1 \times R2}{R1 + R2} = \frac{\text{product of individual resistances}}{\text{sum of individual resistances}}$$

(NB: Appendix C can be used to determine series and parallel resistances *without* using formulae.)

Example 2.9
A resistance of 400kΩ is required. What series combination of resistors will satisfy this requirement?

A 180kΩ and a 220kΩ resistor may be wired in series to provide a resistance of 400kΩ, as shown below:

$$R = R1 + R2 = 180k + 220k = 400k\Omega$$

Both 180k and 220k resistors are commonly available in the E12 and E24 series (note that we have assumed that power rating is unimportant in this example).

Example 2.10
A resistance of 5kΩ (rated at 1W) is required. What parallel combination of resistors will satisfy this requirement?

Two 10kΩ resistors wired in parallel will provide a resistance of 5kΩ, as shown below:

$$R = \frac{R1 \times R2}{R1 + R2} = \frac{10k \times 10k}{10k + 10k} = \frac{100k}{20k} = 5k\Omega$$

Since the resistors are of equal value, they will both dissipate the same power. Hence in order for the combination to be rated at 1W, each individual resistor should be rated at 0.5W. In general, whenever two resistors of identical value are wired in parallel, the resistance of the combination will be equal to *half* that of each individual resistor, whilst the power rating of the combination will be *double* that of each individual component.

The voltage divider

A common application of resistors is that of providing a fixed division of potential using a voltage divider of the form shown in Figure 2.6. The input voltage is divided by a factor which is

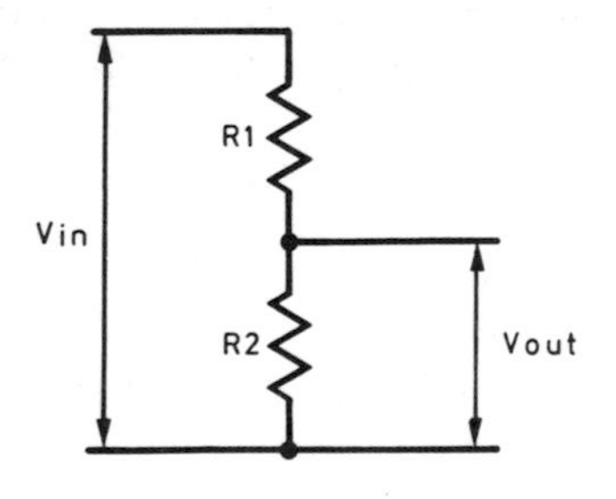

Figure 2.6 Voltage divider

determined by the values of resistor present. The output voltage produced by the circuit of Figure 2.6 is given by:

$$V_{out} = V_{in} \times \frac{R2}{R1 + R2}$$

Example 2.11

An item of test equipment requires an accurate 1V d.c. test voltage. Design a suitable voltage divider arrangement which can make use of an existing regulated 5V d.c. supply rail. (Assume that the input resistance of the equipment connected to the test point is greater than 100kΩ.)

The required voltage division is 1V/5V = 0.2 (or 20%).

Hence $R2/(R1 + R2) = 0.2$

or $R2 = 0.2\,(R1 + R2)$

$R2 = 0.2\,R1 + 0.2\,R2$

Hence $R2 - 0.2\,R2 = 0.2\,R1$

$0.8\,R2 = 0.2\,R1$

$$R2 = \frac{0.2\,R1}{0.8}$$

Thus $R2 = R1/4$

or $R1 = 4\,R2$

Taking a 'worst case' value for the load of 100kΩ, this would suggest that R2 should be 1kΩ and R1 should be four times this value (i.e. 4kΩ). This value can be achieved by connecting a 1.8kΩ resistor in a series with a 2.2kΩ resistor. All resistors should ideally be ±1% 0.25W metal film types. Note that the total resistance seen at the input of the voltage divider will be 5kΩ (i.e. 2.2kΩ + 1.8kΩ + 1kΩ) and it will thus place a 1mA demand on the 5V supply.

Hints and tips

★ In order to obtain accurate values of voltage division, close tolerance (e.g. ±1%) resistors should be used.

★ A notable disadvantage of simple voltage dividers is that the output voltage (V_{out}) will fall when current is drawn from the arrangement and thus accuracy is impaired when such an arrangement is loaded.

★ In order to minimise the effects of an imperfect load, it is important to ensure that the resistance of the circuit to which the voltage divider is connected (ideally infinite) is at least ten times the value of R2. When selecting a value for R2 in precision applications it is wise to choose a value which represents 1% (or less) of the expected load.

★ It is important to ensure that the current demanded from the input voltage source is not excessive (R1 and R2 appear in series across the input voltage). In most applications the input current should not exceed between 1mA and 10mA.

The current divider

Resistors may also be used to divert a known proportion of the current from one branch of a circuit to another. In this case, the input current is divided by a factor which is determined by the

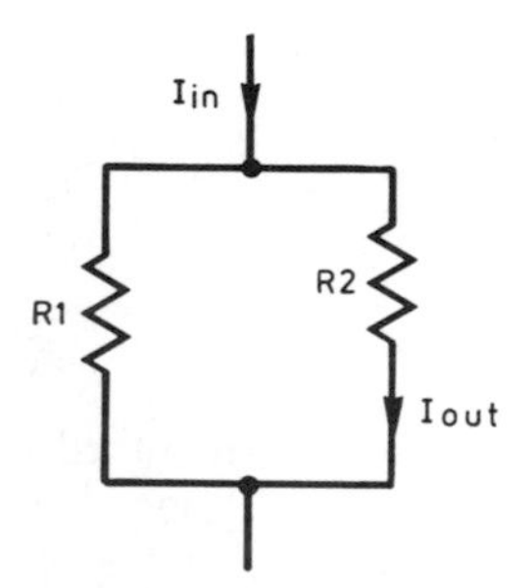

Figure 2.7 Current divider

resistor values present. The current produced by the circuit of Figure 2.7 is given by:

$$I_{out} = I_{in} \times \frac{R1}{R1 + R2}$$

Example 2.12
A moving coil meter operates with a 1mA d.c. full-scale deflection current. If the meter has a coil resistance of 100Ω, devise a suitable current divider arrangement which will provide full-scale deflection when 10mA is applied. Calculate the required value of parallel (shunt) resistor.

The required current division is 1mA/5mA = 0.2 (or 20%). Hence R1/(R1+R2) = 0.2.

The value of R2 is simply that of the meter coil (i.e. 100Ω) as shown in Figure 2.8.

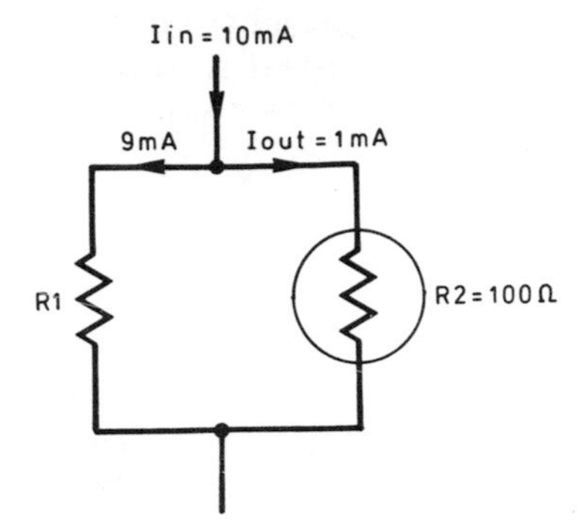

Figure 2.8 See Example 2.12

$$\text{Now } \frac{R1}{R1 + 100} = 0.2$$

or $R1 = 0.2\,(R1 + 100)$

$R1 = 0.2R1 + 20$

Thus $R1 - 0.2\,R1 = 20$

$0.8\,R1 = 20$

or $R1 = 20/8 = 2.5\Omega$

The required value of R1 can be achieved by connecting 3.3Ω and 10Ω resistors in parallel across the terminal of the meter. The resistors used for R1 should ideally be ±2% 0.25W thick film metal glaze resistors. Note that, in this case, R2 is the internal resistance of the meter – not an individual discrete component!

Hints and tips

★ In order to obtain accurate values of current division, close tolerance (e.g. ±1%) resistors should be used.

★ A notable disadvantage of simple current dividers is that the output current (I_{out}) will fall, when the load connected between the output terminals has any appreciable resistance, thus impairing accuracy.

★ A general rule of thumb is that the resistance of the load (ideally zero) should be no more than one tenth of the value of R2.

★ In order to minimize effects of an imperfect load, it is important to ensure that the resistance of the circuit to which the current divider is connected is one tenth (or less) than the value of R2.

Furthermore, when selecting a value for R2 in precision applications it is wise to choose a value which is at least 100 times greater than the expected load resistance.

Kirchhoff's laws

Kirchhoff's laws relate to the algebraic sum of currents at a junction (or *node*) or voltages in a network (or *mesh*). The term *algebraic* simply indicates that the polarity of each current or voltage drop must be taken into account by giving it an appropriate sign, either positive (+) or negative (−).

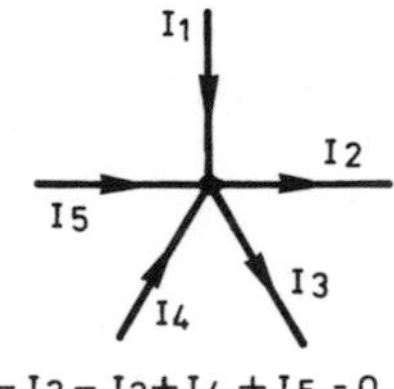

CONVENTION:
CURRENTS FLOWING TOWARDS THE JUNCTION ARE POSITIVE (+)
CURRENTS FLOWING AWAY FROM THE JUNCTION ARE NEGATIVE (−)

(a)

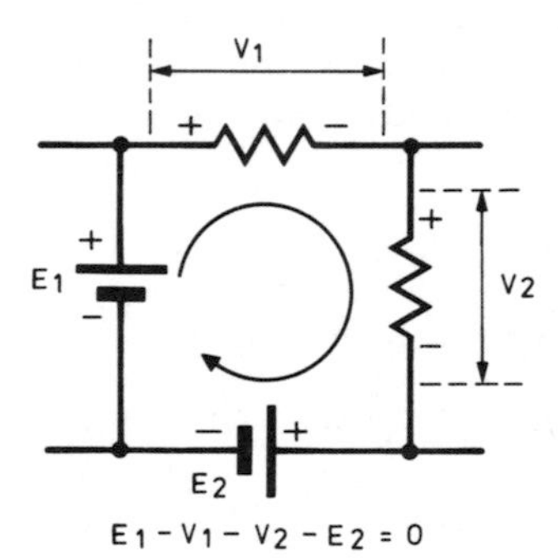

CONVENTION:
CLOCKWISE E.M.F.'S AND POTENTIAL DROPS ARE POSITIVE (+)
ANTICLOCKWISE E.M.F.'S AND POTENTIAL DROPS ARE NEGATIVE (−)

(b)

Figure 2.9 Kirchoff's laws: (a) Kirchoff's current law; (b) Kirchoff's voltage law

Kirchhoff's current law states that the algebraic sum of the currents present at a junction (node) in a circuit is zero (see Figure 2.9(a)).

Kirchhoff's voltage law states that the algebraic sum of potential drops in a closed network (mesh) is zero (see Figure 2.9(b)).

Example 2.13

A 12V supply rated at 500mA (see Figure 2.10) delivers the following currents:

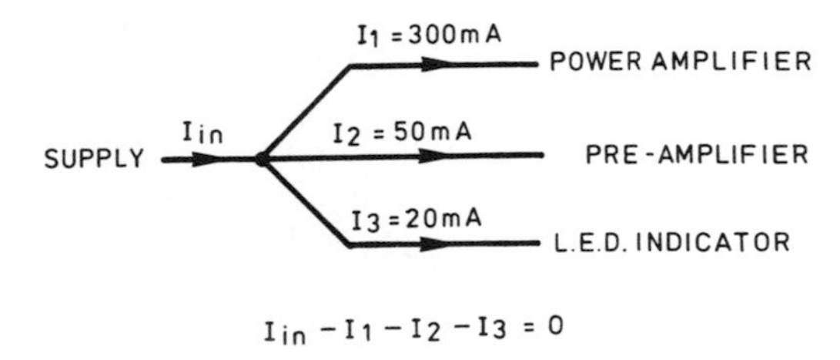

Figure 2.10 See Example 2.13

300mA for a power amplifier output stage
50mA for a pre-amplifier stage
20mA for an LED indicator

What is the total current supplied and how much reserve current is available?

Kirchhoff's current law may be applied, as shown in Figure 2.10. The total current demand is found simply by adding the currents ($I_{IN} = I_1 + I_2 + I_3$ since $I_{IN} - I_1 - I_2 - I_3 = 0$). Thus:

$I_{IN} = 300\text{mA} + 50\text{mA} + 20\text{mA} = 390\text{mA}$

The total current supplied will thus be 390mA. Since the supply is rated at 500mA, a reserve of 500mA − 390mA = 110mA will be available.

Example 2.14

Part of the supply circuitry for a portable radio is shown in Figure 2.11. Determine a suitable value for R.

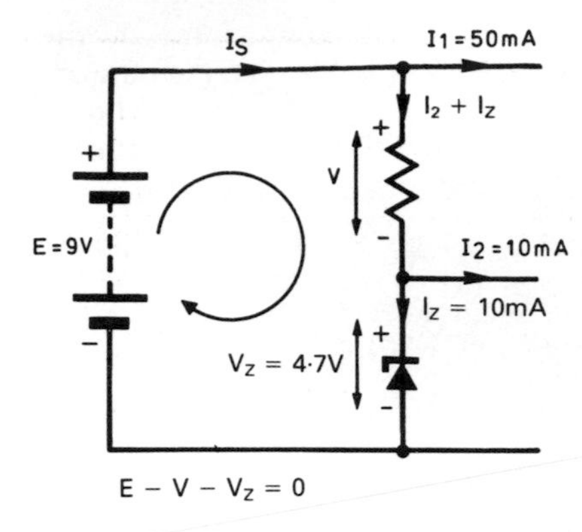

Figure 2.11 See Example 2.14

From Kirchhoff's current law: $I_s - I_1 - I_2 - I_z = 0$
Therefore $I_s = I_1 + I_2 + I_z$
or $I_s = 50mA + 10mA + 10mA$
Thus $I_s = 70mA$

From Kirchhoff's voltage law: $E - V - V_z = 0$
Therefore $V = E - V_z$
or $V = 9V - 4.7V$
Thus $V = 4.3V$

From Ohm's law: $R = V/I = V/(I_2 + I_z) = 4.3V/20mA = 215\Omega$.

In practice, the nearest preferred value (220Ω) would be perfectly adequate.

Preset resistors

Preset resistors allow adjustments to be carried out on electronic circuits without the need to make changes to fixed resistors. Various forms of preset resistor are commonly available including open carbon track skeleton presets (for both horizontal and vertical PCB mounting) and fully encapsulated carbon and multi-turn cermet types. The following table should assist readers when selecting preset resistors:

	Preset resistor type				
	Open skeleton carbon	*Enclosed carbon*	*Open cermet*	*Enclosed cermet*	*Multi-turn cermet*
Resistance range (Ω)	100 to 2.2M	470 to 1M	100 to 1M	100 to 1M	100 to 1M
Typical tolerance (%)	±20	±20	±20	±10	±10
Power rating (W)	0.2	0.15	0.75	0.5	0.25
Temperature coefficient (ppm/°C)	−500 typical	−500 typical	−125 to +200	±100	±100
Stability	poor	poor	fair	good	good
Noise performance	very poor	poor	fair	good	good
Ambient temperature range (°C)	−55 to +125	−55 to +125	−40 to +125	−55 to +125	−55 to +125
Typical unit cost	15p	15p	40p	60p	£1
Typical applications		general purpose		small-signal amplifiers, test and measuring equipment	

Variable resistors

Like preset resistors, variable resistors are available in variety of forms, most popular of which are the carbon track and wirewound *potentiometer* (a three-terminal variable resistor). Carbon potentiometers are available with linear and semi-logarithmic law tracks and in rotary or linear/slider formats.

Controls are also available in which several carbon track potentiometers are linked together by a common control shaft. These components are often referred to as *ganged, tandem* or *stereo* controls. The following table should assist readers when selecting variable resistors:

	Variable resistor type			
	Carbon track	*Cermet*	*Wirewound*	*Multiturn wirewound*
Resistance range (Ω)	5k to 1M	10 to 1M	10 to 100k	100 to 100k
Typical tolerance (%)	±20	±10	±5	±5
Power rating (W)	0.25	1 to 5	1 to 3	1.5 to 3
Temperature coefficient (ppm/°C)	−500 typical	±150	+50 typical	+50 typical
Stability	poor	good	good	good
Noise performance	poor	good	good	good
Ambient temperature range (°C)	−10 to +70	−40 to +85	−20 to +100	−55 to +125
Typical unit cost	50p	£2	£1.50	£4
Typical applications	general purpose	power supplies, test and measuring equipment		Test and measuring equipment

Hints and tips

★ Avoid using open carbon track preset resistors – they are comparatively noisy and unreliable. Cermet components should be used whenever possible.

★ Carbon track variable resistors suffer from noise and track wear. This makes them unacceptable for critical applications (such as instrumentation, low-noise amplifiers etc).

★ To give better control over output levels in audio applications, use logarithmic law potentiometers for volume controls.

★ Carbon track controls should never be used for performing adjustments of output voltage in power supplies as intermittent slider contacts can sometimes result in the full output voltage being applied momentarily to the load!

Thermistors and varistors

Unlike conventional resistors, the resistance of a *thermistor* is intended to change considerably with temperature. Thermistors are thus employed in temperature sensing and temperature compensating applications. Two basic types of thermistor are available; negative temperature coefficient (*n.t.c.*) and positive temperature coefficient (*p.t.c.*). Typical n.t.c. thermistors have resistances which vary from a few hundred (or thousand) ohms at 25°C to a few tens (or hundreds) of ohms at 100°C (see Figure 2.12). P.t.c. thermistors, on the other hand, usually have a resistance-temperature characteristic which remains substantially flat (typically at around 100Ω) over the range 0°C to around 75°C. Above this, and at a critical temperature (usually in the range 80°C to 120°C) their resistance rises very rapidly to values of up to, and beyond, 10kΩ (see Figure 2.13).

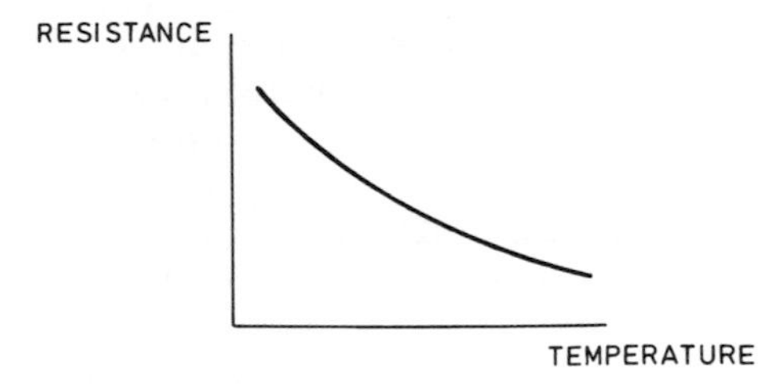

Figure 2.12 N.T.C. thermistor characteristics

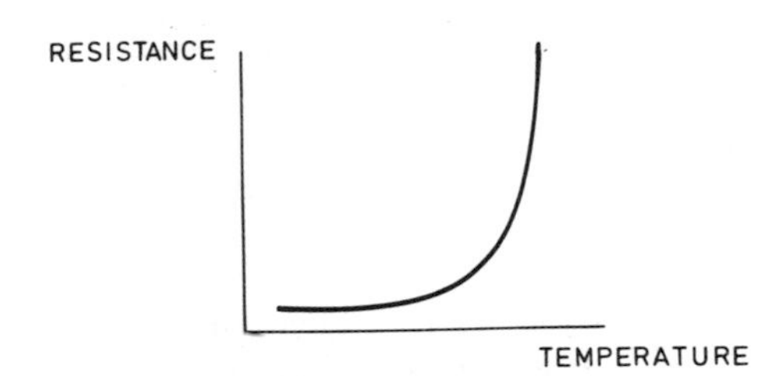

Figure 2.13 P.T.C. thermistor characteristics

A typical application of p.t.c. thermistors is over-current protection. Provided the current passing through the thermistor remains below the threshold current, the effects of self heating will remain negligible and the resistance of the thermistor will remain low (i.e. approximately the same as the resistance quoted at 25°C). Under fault conditions, the current exceeds the threshold value and the thermistor starts to self heat. The resistance then increases

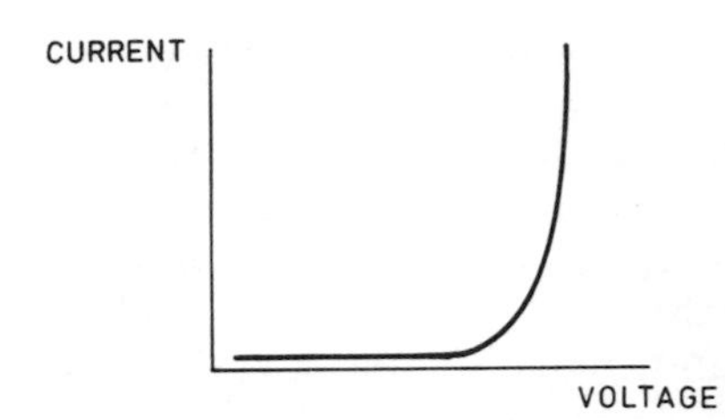

Figure 2.14 Voltage dependent resistor characteristics

rapidly and the current falls to the rest value. Typical values of threshold and rest currents are 200mA and 8mA respectively for a device which exhibits a nominal resistance of 25Ω at 25°C.

Voltage dependent resistors (*VDR*) or *varistors* exhibit a resistance which decreases as the applied voltage increases (see Figure 2.14). Varistors are often used to suppress mains borne transients and voltage spikes, which are produced when inductive loads are switched. The following data refers to a range of commonly available varistors:

Varistor type	*Nominal voltage (V)*	*Transient energy absorption (J)*	*Peak current (A)*	*Continuous voltage (V_{rms})*	*(V_{dc})*	*Max. clamping voltage V_c (V)*	*at I_p (A)*	*Typical capacitance at 1MHz (F)*
V22ZA1	22	0.9	250	14	18	47	5	1.6n
V33ZA5	33	6.0	1k	20	26	64	10	6n
V100ZA3	100	5.0	1.2k	60	81	165	10	400
V130LA5	200	20	2.5k	130	175	340	25	450
V250LA4	390	21	1.2k	250	330	650	10	90
V250LA10	390	40	2.5k	250	330	650	25	220
V275LA4	430	23	1.2k	275	369	710	10	80
V275LA10	430	45	2.5k	275	369	710	25	200
V420LA10	680	45	2.5k	420	560	1.1k	25	140

Capacitors

Capacitors provide us with a means of storing electrical energy in an *electric field*. Typical applications include energy reservoirs and smoothing circuits in power supplies, coupling a.c. signals between the stages of amplifiers, and decoupling supply rails (i.e.

effectively grounding the supply rails to residual a.c. signals and noise).

The electrical characteristics of a capacitor are determined by a number of factors including the dielectric material and physical dimensions. Factors which must be considered when selecting a capacitor for a particular application include:

(a) The required *value* of capacitance (expressed in μF, nF, or pF)
(b) The required *voltage rating* (i.e. the maximum voltage which can be continuously applied to the capacitor under a given set of conditions). It is important to note that applied voltages which are in excess of the rated voltage may have dire consequences as the dielectric may rupture and the capacitor may suffer irreversible damage.
(c) The desired *accuracy* or *tolerance* (quoted as the maximum permissible percentage deviation from the marked value).
(d) The *temperature coefficient* of the capacitance (usually expressed in parts per million, ppm, per unit temperature change).
(e) The *stability* of the capacitor. Usually expressed in terms of the long or short-term percentage variation or capacitance which occurs under specified physical and electrical operating conditions.
(f) The *leakage current* (ideally zero) flowing in the dielectric when the rated d.c. voltage is applied (normally quoted at a given temperature). Alternatively, an *insulation resistance* may be specified. This is the resistance measured between the capacitor plates (ideally infinite) under a given set of conditions.

The following table summarises the properties of commonly available types of capacitor:

Capacitor type	*Ceramic*	*Electrolytic*	*Metallised film*
Capacitance range (F)	2.2p to 10n	100n to 68m	1μ to 16μ
Typical tolerance (%)	±10 and ±20	−10 to +50	±20
Typical voltage rating (d.c.)	50V to 250V	6.3V to 400V	250V to 600V
Temperature coefficient (ppm/°C)	+100 to −4700	+1000 typical	+100 to +200

continued overleaf

Capacitor type	*Ceramic*	*Electrolytic*	*Metallised film*
Stability	fair	poor	fair
Ambient temperature range (°C)	−85 to +85	−40 to +85	−25 to +85
Typical unit cost	6p	10p upwards (note 1)	£8 upwards
Typical applications	decoupling, temperature compensation	decoupling, reservoir and smoothing circuits in power supplies	high voltage power supplies (note 2)
Capacitor type	*Mica*	*Polyester*	*Poypropylene*
Capacitance range (F)	2.2p to 10n	10n to 2.2μ	1n to 470n
Typical tolerance (%)	±1	±20	±20
Typical voltage rating (d.c.)	350V	250V	1kV
Temperature coefficient (ppm/°C)	+50	+200	−200
Stability	excellent	good	good
Ambient temperature range (°C)	−40 to +85	−40 to +100	−55 to +100
Typical unit cost	20p	10p to 30p	20p to 50p
Typical applications	tuned circuits, filters, oscillators	general purpose	high-voltage a.c. circuits
Capacitor type	*Polycarbonate*	*Polystyrene*	*Tantalum*
Capacitance range (F)	10n to 10μ	10p to 10n	100n to 100μ
Typical tolerance (%)	±20	±2.5	±20
Typical voltage rating (d.c.)	63V to 630V	160V	6.3V to 35V
Temperature coefficient (ppm/°C)	+60	−150 to +80	+100 to +250

Capacitor type	*Polycarbonate*	*Polystyrene*	*Tantalum*
Stability	excellent	good	fair
Ambient temperature range (°C)	−55 to +100	−40 to +70	−55 to +85
Typical unit cost	30p upwards	10p upwards	15p upwards
Typical applications	filters, oscillators, timing	general purpose timing and decoupling	general purpose, timing and decoupling (note 3)

Note 1 Costs depend very much upon value and voltage rating. Large value aluminium can-type electrolytics can be expensive.
Note 2 Large physical size (unsuitable for PCB mounting).
Note 3 Small physical size (generally very much smaller than comparable tubular electrolytics).

Hints and tips

★ Working voltages are related to operating temperatures, and capacitors should be derated at high temperatures. Where reliability is important capacitors should be operated at well below their nominal maximum working voltages.

★ Where the voltage rating is expressed in terms of a direct voltage (e.g. 250V d.c.) unless otherwise stated, this is related to the maximum working temperature. It is, however, always wise to operate capacitors with a considerable margin for safety which also helps to ensure long term reliability. As a rule of thumb, the working d.c. voltage should be limited to no more than 50% to 60% of the rated d.c. voltage.

★ Where an a.c. voltage rating is specified this is normally for sinusoidal operation at either 50Hz or 60Hz. Performance will not be significantly affected at low frequencies (up to 100kHz, or so) but, above this, or when non-sinusoidal (e.g. pulse) waveforms are involved, the capacitor must be derated in order to minimise dielectric losses which can produce internal heating and lack of stability.

★ Large value electrolytic and metallised film capacitors can retain an appreciable charge for some considerable time. In the case of components operating at high voltages, a carbon film 'bleed'

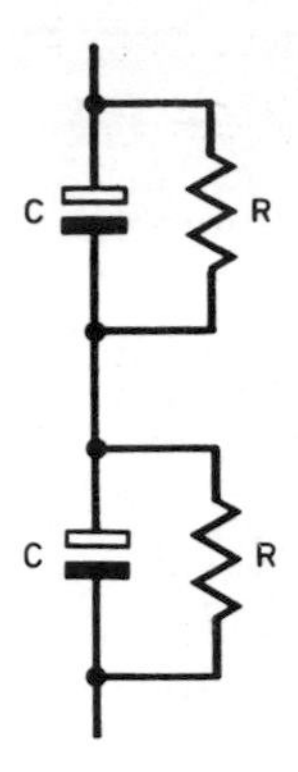

Figure 2.15 Use of resistors to equalise potentials

resistor of typically 1MΩ 0.5W) should be connected in parallel with the capacitor to provide a discharge path.

★ In high-voltage applications, capacitors may be connected in series in order to increase the effective working voltage In such cases, it is necessary to equalise the d.c. polarising potential developed across the capacitor, as shown in Figure 2.15 (a typical value for each of the equalising resistors would be in the range 220kΩ to 1MΩ).

★ Ceramic feed-through capacitors are available for decoupling at very high frequencies (e.g. above 30MHz). Such components may be soldered directly into a bulkhead or screening enclosure and are typically rated at 1nF 350V.

★ Non-polarised electrolytic capacitors are available for applications in which a direct polarising voltage cannot readily be obtained (e.g. loudspeaker cross-over networks). Typical values for non-polarised electrolytics range from 1μF to 100μF (rated at up to 50V r.m.s.). Note, however, that these capacitors are more expensive than conventional (polarised) electrolytics.

★ When selecting a reservoir capacitor for use in a power supply, care must be taken to ensure that the component has an adequate ripple current rating (the typical ripple current rating for a 10000μF can-type electrolytic is often in excess of 5A).

★ Care must be exercised when using electrolytic capacitors for interstage coupling as the leakage current may significantly affect the bias conditions. It is also imperative to ensure that the component is fitted with the correct polarity.

★ Electrolytic capacitors exhibit a fairly wide tolerance and hence in the majority of smoothing and decoupling applications, it is

usually quite permissible to substitute one value for another, provided the working voltage of the substitute capacitor is of the same, or higher, value.
★ The outer foil connection of polystyrene capacitors is often coded with a coloured stripe. For more effective decoupling and to minimise radiation of noise, this connection should be returned to ground or 0V.

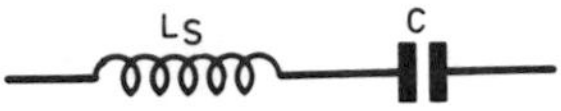

Figure 2.16 Equivalent circuit of a capacitor at high frequencies

★ At high frequencies, stray lead inductance becomes significant. The stray lead inductance (L_s) of the component forms a series resonant circuit with the capacitance (C), as shown in Figure 2.16.

Capacitor colour codes and markings

The vast majority of capacitors employ written markings which indicate their values, working voltages, and tolerance. The notable exception to this rule is the colour code (shown in Figure 2.17)

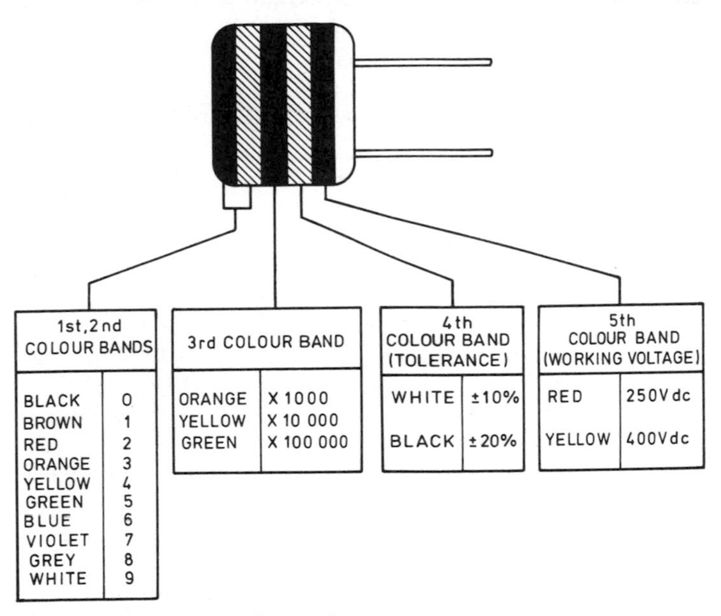

Figure 2.17 Capacitor colour code

which was, until comparatively recently, commonly used to mark the values of resin dipped polyester capacitors.

The more modern method of marking resin dipped polyester (and other) types of capacitor involves quoting the value (in μF, nF or pF), the tolerance (often either ±10% or ±20%), and the working voltage (using ⎓ and ~ to indicate d.c. and a.c. respectively). Several manufacturers use two separate lines for their capacitor markings and these have the following meanings:

First line: capacitance (in pF or μF) and tolerance (K=10%, M=20%).
Second line: rated d.c. voltage and code for the dielectric material.

A three-digit code is commonly used to mark monolithic ceramic capacitors. The first two digits correspond to the first two digits of the value whilst the third digit is a multiplier which gives the number of zeroes to be added to give the value in pF.

Another method of coding is employed to indicate the temperature coefficient of ceramic capacitors which are frequently employed for temperature compensation in the tuned circuits of oscillators and filters. The following system is employed (N denotes negative temperature coefficient):

Marking	*Temperature coefficient (ppm/deg. C)*	*Tip colour*
NP0	0	black
N030	−30	brown
N080	−80	red
N150	−150	orange
N220	−220	yellow
N330	−330	green
N470	−470	blue
N750	−750	violet
N1500	−1500	orange/orange
N2200	−2200	yellow/orange
N3300	−3300	green/orange
N4700	−4700	blue/orange

Example 2.15
A monolithic ceramic capacitor is marked with the legend 103. What is its value?

10000pF or 10nF

Example 2.16

A polyester capacitor is marked with the legend: 0.22/20 250— What is its value, tolerance, and working voltage?

0.22μF, ±20%, 250V d.c.

Series and parallel combinations of capacitors

In order to obtain a particular value of capacitance, fixed capacitors may be arranged in either series or parallel (Figures 2.18 and 2.19).

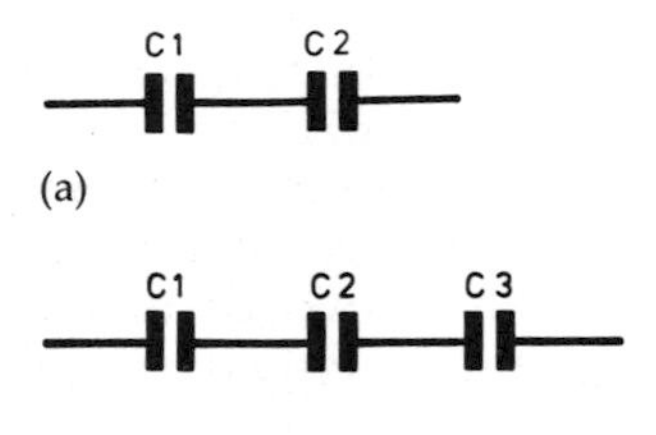

Figure 2.18 Series connected capacitors: (a) two capacitors in series; (b) three capacitors in series

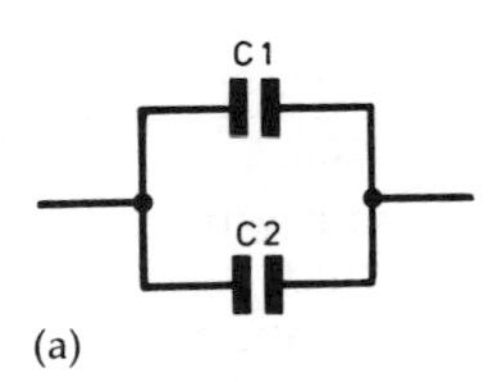

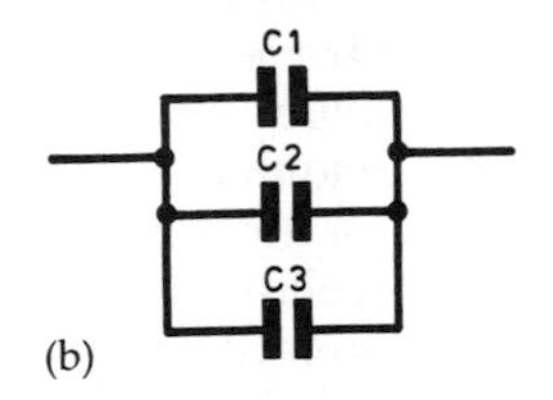

Figure 2.19 Parallel connected capacitors: (a) two capacitors in parallel; (b) three capacitors in parallel

The reciprocal of the effective capacitance of each of the series circuits shown in Figure 2.18 is equal to the sum of the reciprocals of the individual capacitances. Hence, for Figure 2.18(a);

$$\frac{1}{C} = \frac{1}{C1} + \frac{1}{C2}$$

whilst for Figure 2.18(b);

$$\frac{1}{C} = \frac{1}{C1} + \frac{1}{C2} + \frac{1}{C3}$$

In the former case, the formula can be more conveniently rearranged as follows;

$$C = \frac{C1 \times C2}{C1 + C2}$$

(This can be simply remembered as the product of the two capacitor values divided by the sum of the two values). For parallel arrangements of capacitors, the effective capacitance of the network is simply equal to the sum of the individual capacitances. Hence, for Figure 2.19(a);

$$C = C1 + C2$$

Whilst for Figure 2.19(b);

$$C = C1 + C2 + C3$$

(NB: Appendix C can be used to determine series and parallel capacitance without using formulae.)

Example 2.17
A capacitance of 3.2μF is required. What parallel combination of capacitors will satisfy this requirement?

A 1μF and a 2.2μF capacitor may be wired in parallel to provide a capacitance of 3.2μF, as shown below:

$$C = C1 + C2 = 1\mu + 2.2\mu = 3.2\mu F$$

Both 1μF and 2.2μF capacitors are commonly available. Note that both capacitors should have similar voltage ratings and, if electrolytic, they should be connected with the same polarity (positive to positive and negative to negative).

Example 2.18
A capacitance of 50μF (rated at 100V) is required. What series combination of capacitors will satisfy this requirement?

Two 100μF capacitors wired in series will provide a capacitance of 50μF, as shown below:

$$C = \frac{C1 \times C2}{C1 + C2} = \frac{100\mu \times 100\mu}{100\mu + 100\mu} = \frac{100\mu}{20\mu} = 50\mu$$

Since the capacitors are of equal value, the applied d.c. potential will be shared equally between them. However, in order to *ensure* that this is the case, a pair of equalising resistors (as shown in Figure 2.15, should be employed). Each capacitor should be rated at more than 50V (63V would be adequate).

Capacitive reactance

The reactance of a capacitor is defined as the ratio of applied voltage to current and, like resistance, it is measured in Ω. The reactance of a capacitor is inversely proportional to both the value of capacitance and the frequency of the applied voltage. Capacitive reactance can be found by applying the following formula:

$$X_c = \frac{V_c}{I_c} = \frac{1}{2\pi f C}$$

where X_c is the reactance in Ω, f is the frequency in Hz, and C is the capacitance in F. Capacitive reactance falls as frequency increases, as shown in Figure 2.20.

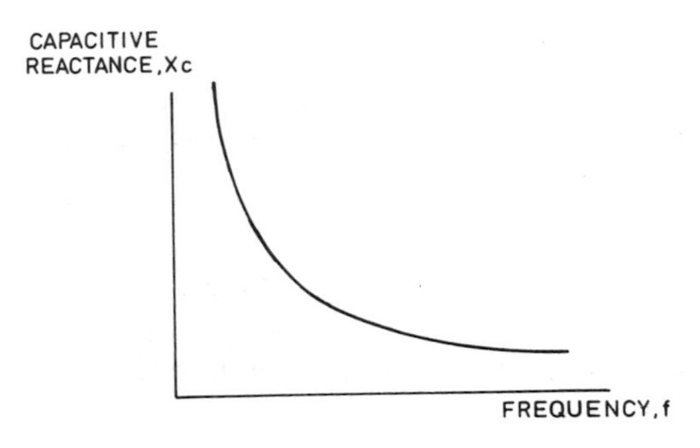

Figure 2.20 Variation of capacitive reactance with frequency

The applied voltage, V_c, and current, I_c, flowing in a pure capacitive reactance will differ in phase by an angle of 90° or $\pi/2$ radians (the current *leads* the voltage). This relationship is illustrated in the current and voltage waveforms (drawn to a common time scale) shown in Figure 2.21, and as a *phasor* diagram shown in Figure 2.22.

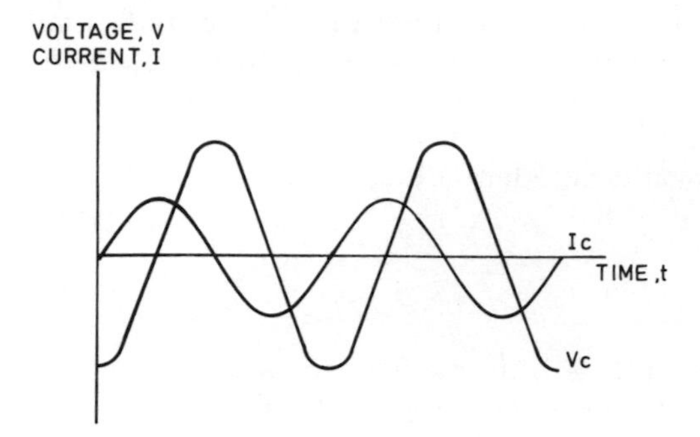

Figure 2.21 Waveforms of current and voltage in a pure capacitive reactance

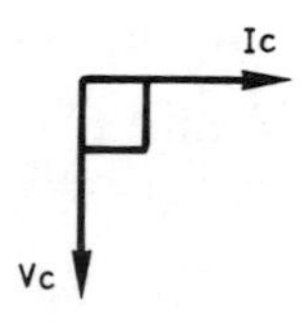

Figure 2.22 Phasor representation of Figure 2.21

Example 2.17
Determine the reactance of a 1μF capacitor at (a) 100Hz and (b) 10kHz.

(a) At 100Hz, $X_c = \dfrac{1}{2 \times \pi \times 100 \times 1 \times 10^{-6}}$

or $\quad X_c = \dfrac{0.159}{10^{-4}} = 0.159 \times 10^4$

Thus $\quad X_c = 1.59\text{k}\Omega$

(b) At 10kHz, $X_c = \dfrac{1}{2 \times \pi \times 10000 \times 1 \times 10^{-6}}$

or $\quad X_c = \dfrac{0.159}{10^{-2}} = 0.159 \times 10^2$

Thus $\quad X_c = 15.9\Omega$

Example 2.14
A 100nF capacitor is to form part of a filter connected across a 240V 50Hz mains supply. What current will flow in the capacitor?

The reactance of the capacitor will be given by:

$$X_c = \frac{1}{2 \times \pi \times 50 \times 100 \times 10^{-9}} = 31.8\text{k}\Omega$$

The r.m.s. current flowing in the capacitor will be:

$$I_c = \frac{V_c}{X_c} = \frac{240\text{V}}{31.8\text{k}\Omega} = 7.5\text{mA}$$

Inductors

Inductors provide us with a means of storing electrical energy in a *magnetic field*. Typical applications include chokes, filters, and frequency selective circuits. The electrical characteristics of an inductor are determined by a number of factors including the material of the core (if any), the number of turns, and the physical dimensions of the coil. Factors which must usually be considered when selecting an inductor for a particular application include:

(a) The required *value* of inductance (expressed in H, mH, μH, or nH).
(b) The required *current rating* (i.e. the maximum current which can be continuously applied to the inductor under a given set of conditions). It is important to note that applied currents which are in excess of the rated voltage may have dire consequences as the coil windings may overheat and fuse causing irreversible damage. Furthermore, the inductance of an iron cored inductor tends to fall as the applied direct current increases. In extreme cases the core may become saturated with magnetic flux and the inductance may fall to an unacceptably low value.
(c) The desired *accuracy* or *tolerance* (quoted as the maximum permissible percentage deviation from the marked value).
(d) The *temperature coefficient* of the inductance (usually expressed in parts per million, ppm, per unit temperature change).
(e) The *stability* of the inductor. Usually expressed in terms of the long or short-term percentage variation of capacitance which occurs under specified physical and electrical operating conditions.
(f) The *d.c. resistance* of the coil windings (ideally zero).
(g) The *Q-factor* (quality factor) of the coil (usually stated at a particular working frequency). Q-factor is dependent upon the value of inductance and the losses of the inductor (which are, in turn, governed by the combined a.c. and d.c. *loss resistance* of the

component. The higher the Q-factor, the better the quality of the inductor.
(h) The *frequency range* for the component (usually stated as upper and lower working frequency limits).

The following table summarises the properties of commonly available types of inductor:

Inductor type	*Single-layer, open*		*Multi-layer, open*		*Multi-layer, pot cored*	*Multi-layer, iron cored*
Core material	air	ferrite	air	ferrite	ferrite	iron
Inductance range (H)	50n to 10μ	1μ to 100μ	5μ to 500μ	10μ to 1m	1m to 100m	20m to 20H
Typical d.c. resistance (Ω)	0.05 to 1	0.1 to 10	1 to 20	2 to 100	2 to 100	10 to 400
Typical tolerance (%)	±10	±10	±10	±10	±10	±10
Typical Q-factor	60	80	100	80	40	20
Typical current rating (A) (note 1)	0.1	0.1	0.2	0.5	0.5	0.1
Typical frequency range (Hz)	5M to 500M	1M to 200M	200k to 20M	100k to 10M	1k to 1M	50 to 10k
Typical unit cost	30p	40p	40p	60p	£1.50	£5
Typical applications	tuned circuits		filters and HF transformers		LF and MF filters and transformers	smoothing chokes and LF filters

Note 1 Depends upon inductance value – larger inductance values generally have lower current ratings.

Several manufacturers can supply a range of fixed radio frequency inductors in preferred values in the E6 series between 1μH and 10mH. In many applications (e.g. when the inductor is required to form part of a resonant circuit in conjunction with a fixed series or parallel connected capacitor) the inductance value should be adjustable to allow precise tuning of the arrangement. In such cases readers can make use of a range of adjustable inductors produced by manufacturers (such as Toko) or wind an appropriate number of turns on a suitable coil former fitted with an adjustable ferrite core.

Inductor formulae

The following *approximate* formulae should assist readers with the task of constructing an inductor:

Single-layer coils where length:diameter ratio exceeds 5:1

$$L = \frac{0.2\ N^2\ d^2}{20l + 9d} \quad \text{and} \quad N = \frac{2.24}{d}\sqrt{L\ (20l + 9d)}$$

where L is the inductance value (in μH), N is the number of turns, d is the diameter of the former (in cm), and l is the length of the coil winding (in cm).

Multi-layer coils where diameter:length ratio exceeds 1:1

$$L = \frac{N^2\ {d_m}^2}{100\ d} \quad \text{and} \quad N = \frac{10}{d_m}\sqrt{L\ d}$$

where L is the inductance value (in μH), N is the number of turns, d_m is the mean diameter of the winding (in cm), and d is the depth of winding (in cm).

Single and multi-layer coils with fixed or adjustable ferrite cores will have inductances *increased* by a factor of between 1.5 and 3 depending upon the grade of ferrite and its position. Brass cores may be fitted (in place of ferrite cores) to *reduce* the inductance values by factors of between 0.6 and 0.9 depending upon position.

Example 2.16

An air cored inductor of value 40μH is required to form part of an RF filter. If a former of diameter 0.8cm and length 5cm is available, determine the number of turns required.

Since the ratio of length:diameter exceeds 5:1 we shall use the formula:

$$N = \frac{2.24}{d}\sqrt{L\ (20l + 9d)}$$

where $d = 0.8\text{cm}$, $l = 5\text{cm}$, and $L = 40\mu H$

$$\text{Hence } N = \frac{2.24}{0.8}\sqrt{40\,((20 \times 5) + (9 \times 0.8))}$$

$$\text{thus} \quad N = 2.8\sqrt{40\,(100 + 7.2)}$$

$$\text{or} \quad N = 2.8\sqrt{40 \times 107.2} = 2.8\sqrt{4288} = 2.8 \times 65.5$$

Hence N = 131 turns.

Note that, if it is essential to obtain an exact value of inductance, it is advisable to reduce the number of turns (by approximately 30%) and fit an adjustable ferrite dust core.

At medium and low frequencies or when an appreciable value of inductance is required (i.e. between 100μH and 100mH), inductors can be readily manufactured using one of the range of RM series ferrite pot cores. The core material of these inductors is commonly available in three grades: A13, Q3 and N28, and the complete pot core assembly (comprising a matched pair of core halves, a single-section bobbin, a pair of retaining clips, and a core adjuster) may be purchased in kit form.

The following data refers to the RM series of ferrite cores:

	Core type				
	RM6	*RM6*	*RM7*	*RM10*	*RM10*
Inductance factor, A_L (nH/turn)	160	250	250	250	400
Tolerance	±3%	±3%	±3%	±2%	±3%
Turns factor (turns for 1mH)	79	63.3	63.3	63.3	50
Adjustment range	±20%	±14%	±15%	±17%	±20%
Effective permeability, μ	100	155	146	90	145
Temperature coefficient (ppm/°C)	51 to 154	80 to 241	73 to 219	50 to 149	80 to 239
Frequency range (kHz)	6 to 800	4 to 700	3 to 650	2 to 650	1 to 500
Saturation flux density (mT)	250	250	250	250	250
Maximum turns on bobbin					
(using 0.2mm dia. wire)	205	205	306	612	612
(using 0.5mm dia. wire)	36	36	50	98	98
(using 1.0mm dia. wire)	9	9	11	25	25

The following formula may be used to determine the number of turns necessary for a particular inductance:

$$L = N^2 A_L \quad \text{and} \quad N = \sqrt{\frac{L}{A_L}}$$

where N = number of turns
L = inductance value (in nH)
A_L = inductance factor (in nH/turn).

Note that it is necessary to take into account the number of turns of wire which can be accommodated on the bobbin bearing in mind that the resistance of the winding will increase as the diameter of the wire decreases. To avoid appreciable temperature rise due to I^2R losses within the winding, wires of 0.2mm, 0.5mm and 1mm diameter should be used for r.m.s. currents not exceeding 100mA, 750mA and 4A respectively.

Example 2.18
A 2mH (2000000nH) inductor is required to operate in a filter with an r.m.s. current of approximately 500mA at a frequency of 3.5kHz. Which RM core should be used and how many turns will be required?

We must first eliminate any of the RM cores for which the operating frequency lies outside the recommended working range. On the basis of this consideration (using the foregoing table of data) we would exclude the RM6 core. Secondly we must select an appropriate wire diameter. A wire diameter of 0.5mm will be adequate for an r.m.s. current of 500mA. Thirdly, the number of turns required should be calculated using the formula:

$$N = \sqrt{\frac{L}{A_L}}$$

where L = 2000000 and A_L = 250 (allowing the use of either an RM7 or an RM10 core),

$$N = \sqrt{\frac{2000000}{250}} = 89 \text{ turns}$$

Finally, we should check that the required number of turns (89) of 0.5mm diameter wire can be accommodated on the bobbin. The *only* suitable type of former for this would be the RM10 (which can accommodate up to 98 turns of 0.5mm wire).

Hints and tips

★ Inductance values of iron cored inductors are very much dependent upon the applied direct current and tend to fall rapidly as the value of applied direct current increases. It is important to take this into account when designing filter circuits for power supply and LF applications.

★ Maximum current ratings for inductors are related to operating temperatures and should be derated when high ambient temperatures are expected. Where reliability is important, inductors should be operated at well below their nominal maximum current ratings.

★ Ferrite toroids and beads can be used for effective filtering and as broadband transformers at frequencies above 30MHz. Inductors can be realised very easily using these cores with just a few turns of wire.

★ The stray magnetic field generated by an inductor (particularly an open type) will pervade the space which surrounds it. In some applications (particularly where RF signals are concerned) it is necessary to position inductors in a way that minimises inductive coupling.

★ Like capacitors, inductors also have a self-resonant frequency. The stray parallel capacitance (C_s) forms a parallel resonant circuit

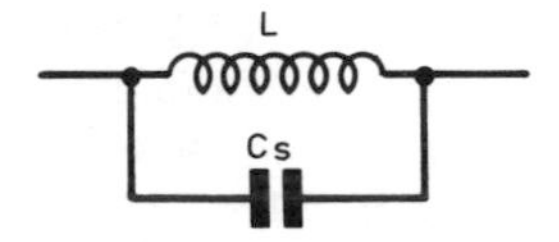

Figure 2.23 Equivalent circuit of an inductor at high frequencies

with the inductance (L) as shown in Figure 2.23. The self-resonant frequency thus depends upon the values of L and C_s. Typical values of self-resonant frequency range from 20kHz for an iron cored choke of appreciable inductance to around 100MHz for a low inductance ferrite cored HF inductor.

Series and parallel combinations of inductors

In order to obtain a particular value of inductance, fixed inductors may be arranged in either series or parallel as shown in Figures 2.24 and 2.25. The effective inductance of each of the series circuits shown in Figure 2.24 are simply equal to the sum of the individual inductances. Hence, for Figure 2.24(a):

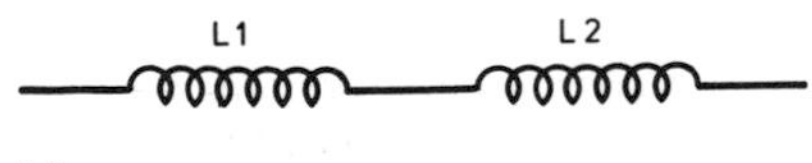

(a)

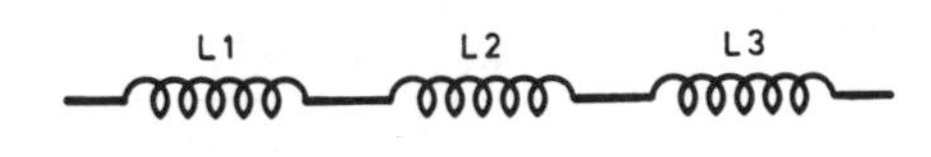

(b)

Figure 2.24 Series connected inductors: (a) two inductors in series; (b) three inductors in series

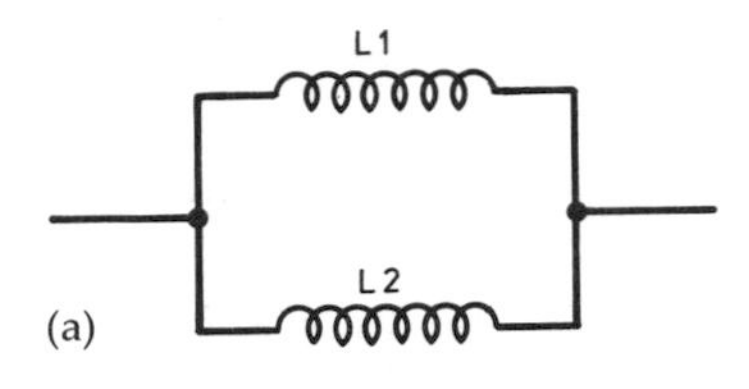

(a)

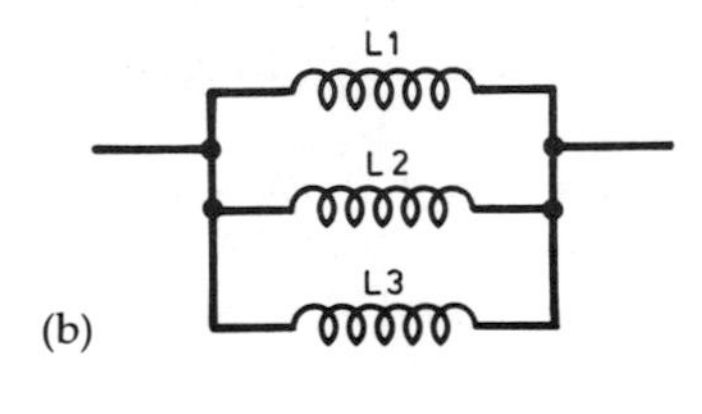

(b)

Figure 2.25 Parallel connected inductors: (a) two inductors in parallel; (b) three inductors in parallel

$$L = L1 + L2$$

Whilst for Figure 2.24(b):

$$L = L1 + L2 + L3$$

Turning to the parallel inductors shown in Figure 2.25, the reciprocal of the effective inductance of each circuit is equal to the sum of the reciprocals of the individual inductances. Hence, for Figure 2.25(a):

$$\frac{1}{L} = \frac{1}{L1} + \frac{1}{L2}$$

whilst for Figure 2.25(b):

$$\frac{1}{L} = \frac{1}{L1} + \frac{1}{L2} + \frac{1}{L3}$$

In the former case, the formula can be more conveniently rearranged as follows;

$$L = \frac{L1 \times L2}{L1 + L2}$$

(This can be simply remembered as the product of the two inductance values divided by the sum of the two values). (NB: Appendix C can be used to determine series and parallel inductance *without* using formulae.)

Example 2.19

An inductance of 5mH is required. What parallel combination of inductors will satisfy this requirement?

Two 10mH inductors may be wired in parallel to provide a inductance of 5mH as shown below:

$$L = \frac{L1 \times L2}{L1 + L2} = \frac{10m \times 10m}{10m + 10m} = \frac{100m}{20m} = 5mH$$

Since the inductors are identical, the applied d.c. current will be shared equally between them. Each inductor should have a current rating which is greater than *half* the current applied to the parallel circuit.

Example 2.20

An inductance of 44mH is required. What series combination of inductors will satisfy this requirement?

Two 22mH capacitors wired in series will provide an inductance of 44mH, as shown below:

$$L = L1 + L2 = 22mH + 22mH = 44mH$$

Since the inductors are wired in series, the applied current will flow through both. Hence they should both be rated at a current which is greater than this value.

Inductive reactance

The reactance of an inductor is defined as the ratio of applied voltage to current and, like resistance, it is measured in Ω. The reactance of an inductor is directly proportional to both the value of inductance and the frequency of the applied voltage. Inductive reactance can be found by applying the formula:

$$X_L = \frac{V_L}{I_L} = 2\,\pi\,f\,L$$

where X_L is the reactance in Ω, f is the frequency in Hz, and L is the inductance in H. Inductive reactance increases linearly with frequency as shown in Figure 2.26.

The applied current, I_L, and voltage, V_L, developed across a pure inductive reactance will differ in phase by an angle of 90° or π/2 radians (the current *lags* the voltage). This relationship is illustrated in the current and voltage waveforms (drawn to a common time scale) shown in Figure 2.27 and as a *phasor* diagram shown in Figure 2.28.

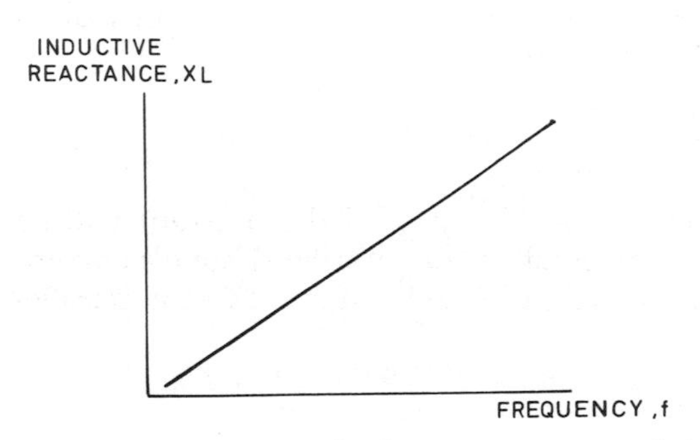

Figure 2.26 Variation of inductive reactance with frequency

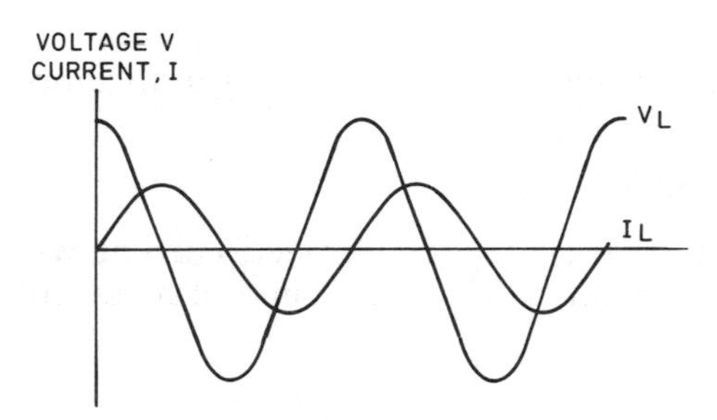

Figure 2.27 Waveforms of current and voltage in a pure inductive reactance

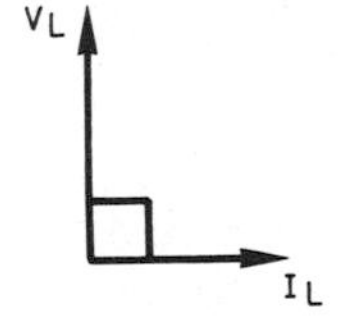

Figure 2.28 Phasor representation of Figure 2.27

Example 2.21

Determine the reactance of a 10mH inductor at (a) 100Hz and (b) at 10kHz.

(a) At 100Hz, $X_L = 2 \times \pi \times 100 \times 10 \times 10^{-3}$
Thus $X_L = 6.28\Omega$
(b) At 10kHz, $X_L = 2 \times \pi \times 10000 \times 10 \times 10^{-3}$
Thus $X_L = 628\Omega$

Example 2.22

A 100mH inductor is to form part of a filter which carries a current of 20mA at 400Hz. What voltage drop will be developed across the inductor?

The reactance of the inductor will be given by:

$X_L = 2 \times \pi \times 400 \times 100 \times 10^{-3} = 251\Omega$

The r.m.s. voltage developed across the inductor will be given by:

$V_L = I_L \times X_L = 20\text{mA} \times 251\Omega = 5.02\text{V}$

In this example, it is important to note that we have assumed that the d.c. resistance of the inductor is negligible by comparison with its reactance. Where this is not the case, it is necessary to determine the *impedance* of the component and use this to determine the voltage drop (see Section 3).

Transformers

Transformers provide us with a means of coupling a.c. power from one circuit to another. The transformation that occurs relates to the voltages present at the *primary* (input) and *secondary* (output) windings. Voltage may be *stepped-up* (secondary voltage greater than primary voltage) or *stepped-down* (secondary voltage less than primary voltage). Since no power gain is provided (in fact, the secondary power will be very slightly less than the primary power due to losses within the transformer) an increase in secondary voltage can only be achieved at the expense of a corresponding reduction in secondary current, and vice versa.

Typical applications include stepping-up or stepping-down mains voltages in power supplies, coupling signals in AF amplifiers to achieve impedance matching, and isolation of d.c. potentials.

The electrical characteristics of a transformer are determined by a number of factors including the core material and physical dimensions. Factors which must be considered when selecting a transformer for a particular application include:

(a) The rated primary and secondary *voltages* and *currents*. A *voltage ratio* or *turns ratio* may also be specified.
(b) The required *power rating* (i.e. the maximum power, usually expressed in volt-amperes, VA) which can be continuously delivered by the transformer under a given set of conditions). It is important to note that loading a transformer beyond its ratings can have dire consequences as the component may overheat causing irreversible damage.
(c) The *frequency range* for the component (usually stated as upper and lower working frequency limits).
(d) The *regulation* of a transformer (usually expressed as a percentage of full-load). The specification is a measure of the ability of a transformer to maintain its rated output voltage under load.

The following table summarises the properties of commonly available types of transformer:

Transformer type	*Ferrite cored*	*Iron cored*	*Iron cored*
Typical power rating	note 1	10mW to 10W	3VA to 500VA
Typical regulation (%)	note 1	note 1	5 to 15
Typical frequency range (Hz)	1k to 100k	50 to 20k	45 to 400
Typical unit cost	£2	£5	£2.50 to £60
Typical applications	pulse applications	AF amplifiers	power supplies

Note 1 Not normally important in transformers of this type.

Several manufacturers can supply a range of mains transformers with various ratings which will meet almost every need. Transformers can also be manufactured on a one-off basis though this can be somewhat epensive and it is well worth shopping around for a stock item before resorting to this. Alternatively, transformers can be constructed from proprietary kits which are available in a variety of sizes (including 20VA, 50VA, 100VA and 200VA) which usually comprise a double-section bobbin complete (with pre-wound mains primary), insulating shrouds, laminations and end mounting frames.

Transformer formulae

The following formulae are useful when constructing low-frequency transformers which use laminated steel or toroidal cores:

Turns ratio and voltage ratio

$$\frac{N_P}{N_s} = \frac{V_P}{V_s}$$

Turns ratio and current ratio

$$\frac{N_P}{N_s} = \frac{I_s}{I_P}$$

Current ratio and voltage ratio

$$\frac{I_P}{I_s} = \frac{V_s}{V_P}$$

where N_P and N_s are the number of primary and secondary turns respectively, V_P and V_s are the primary and secondary voltages respectively, and I_P and I_s are the primary and secondary currents, respectively.

Power rating

The working VA for a power transformer can be estimated by calculating the total power consumed by each secondary load and multiplying this by 1.1 (to allow for losses within the transformer when on load). For a transformer with a *single* secondary winding:

Secondary (load power), $P_L = I_s \times V_s$

Primary (input power), $P_{IN} = 1.1 \times P_L$ (approximately).

Hints and tips

★ Maximum power ratings for transformers are related to operating temperatures and should be derated when high ambient temperatures are expected. Where reliability is important, transformers should be operated at well below their nominal power ratings.

★ The stray magnetic field generated by a transformer will pervade the space which surrounds it. The field around a power transformer will be appreciable and, to avoid stray coupling of signals and *hum*, it may be necessary to exercise care when positioning such a component. For this reason, power transformers should *never* be mounted in close proximity to the early stages of an audio ampifier nor should they be placed adjacent to audio frequency transformers or cathode ray tubes.

★ In order to maintain high efficiency, it is important to avoid saturation within the core of a transformer. If the transformer is being operated under sinusoidal conditions, and provided one keeps within the manufacturer's ratings, saturation should not occur. For pulse operation or when an appreciable direct current is applied to a transformer, it may be necessary to make an estimation of the peak flux density in order to check that saturation is not occurring.

★ When mounting toroidal transformers, care must be taken to ensure that both ends of the fixing bolt do not simultaneously come into contact with a metal chassis or mounting framework as this will cause a *shorted turn* which will greatly reduce efficiency and may cause irreversible damage to the transformer windings.

3 Networks, attenuators and filters

Attenuators

Simple attenuators can be based on nothing more than a potential divider arrangement. Figures 3.1 and 3.2 show typical potential divider arrangements (the latter is based on commonly available preferred values). Each of these circuits has a fixed input resistance (1MΩ for the circuit of Figure 3.1 and 910Ω in the case of Figure 3.2). In both cases, the output (load) circuit must have a comparatively high impedance in order that accuracy is maintained. High stability 1% tolerance resistors should be used in both circuits.

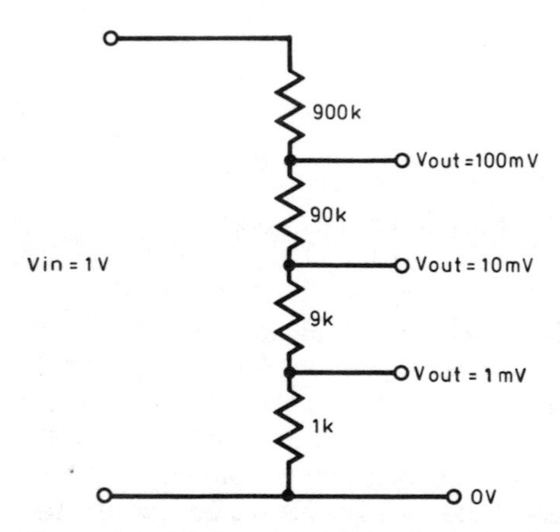

Figure 3.1 Simple potential divider attenuator

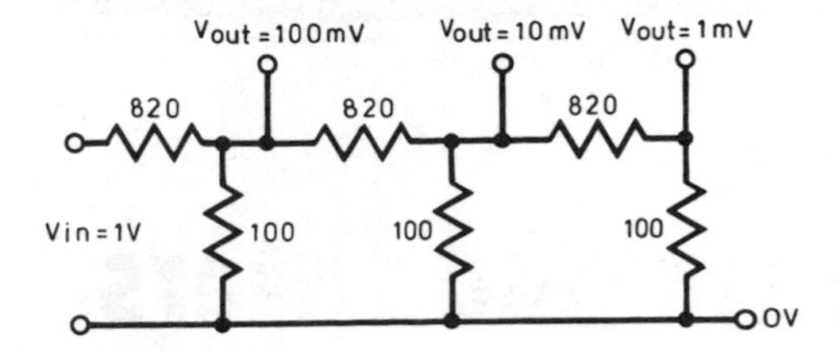

Figure 3.2 Potential divider based on preferred values

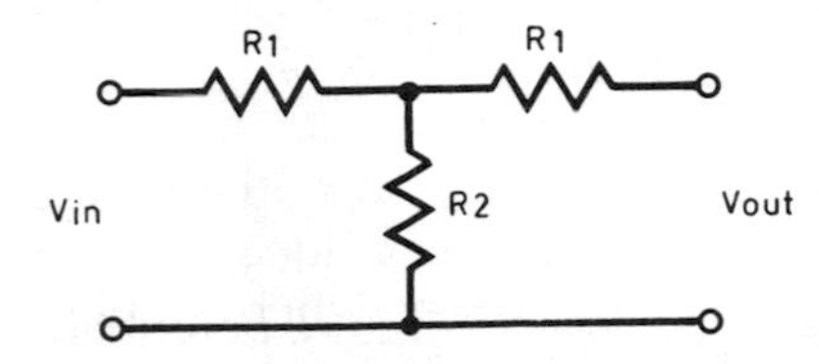

Figure 3.3 T-network attenuator

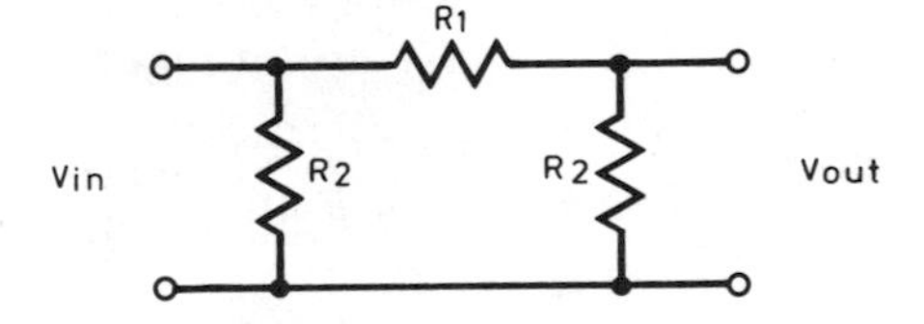

Figure 3.4 π-network attenuator

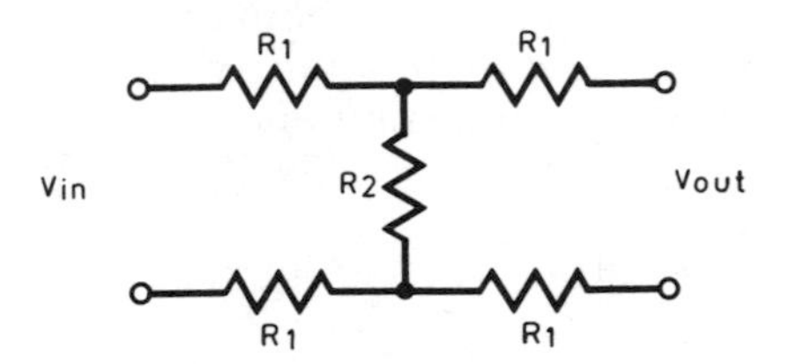

Figure 3.5 H-network attenuator

The simple potential divider type attenuators described in Figures 3.1 and 3.2 do not, unfortunately, possess equal input and output resistances. This is a desirable feature whenever an attenuator is to form part of a matched system. Figures 3.3, 3.4, and 3.5 show typical *constant impedance* attenuators based on T, π and H arrangements of resistors. The design formulae for these circuits are as follows:

Resistor	*Attenuator type* T	π	H
R1	$\frac{Z_o (A-1)}{(A+1)}$	$\frac{Z_o (A^2-1)}{2A}$	$\frac{Z_o (A-1)}{2 (A+1)}$
R2	$\frac{Z_o 2A}{(A^2-1)}$	$\frac{Z_o (A+1)}{(A-1)}$	$\frac{Z_o 2A}{(A^2-1)}$

Where Z_o is the *design impedance* of the attenuator (usually 600Ω for audio and telephone line matching purposes, and 50Ω or 75Ω for video and RF applications) and A is the desired attenuation expressed in terms of the ratio of input to output voltage (i.e. $A = V_{IN}/V_{OUT}$).

The T and π configurations are used for *unbalanced* attenuators in which the *common connection* is effectively grounded. The H configuration is employed in *balanced* attenuators where neither side of the input or output is grounded.

Example 3.1

An unbalanced attenuator is to provide a 10× attenuation of voltage and is to exhibit a constant impedance of 75Ω. Design a suitable arrangement based on preferred value resistors.

Since an unbalanced attenuator is specified we must reject the H configuration which is only suitable for balanced operation. It is then necessary to calculate the resistance values for both the T and π configurations and then employ the configuration which can most easily be realised using preferred value components. If a T configuration is employed.

$$R1 = \frac{75\,(10-1)}{(10+1)} = \frac{75 \times 9}{11} = 61.3\Omega$$

and $$R2 = \frac{75 \times 2 \times 10}{(10^2-1)} = \frac{75 \times 20}{99} = 15.2\Omega$$

If a π configuration is employed:

$$R1 = \frac{75\,(10^2-1)}{2 \times 10} = \frac{75 \times 99}{20} = 371.3\Omega$$

$$R2 = \frac{75\,(10+1)}{(10-1)} = \frac{75 \times 11}{9} = 91.7\Omega$$

The former configuration can be most easily realised using preferred value resistors. A single 15Ω resistor can be used for R2, whilst R1 can be realised using two 120Ω resistors connected in parallel. If resistors of 2% (or better) are employed, the accuracy of the attenuator can be expected to be somewhat better than 4% which will be quite adequate for most applications.

Decibels

Decibels provide us with a convenient means of expressing gain (*amplification*) and loss (*attenuation*) in electronic circuits. The decibel (dB) is one tenth of a bel which, in turn, is defined as the logarithm, to the base 10, of the ratio of output power level (P_{OUT}) to input power level (P_{IN}).

Gain and loss may be expressed in terms of power, voltage and current such that:

$$A_P = \frac{P_{OUT}}{P_{IN}},\ A_V = \frac{V_{OUT}}{V_{IN}} \text{ and } A_I = \frac{I_{OUT}}{I_{IN}}$$

Where A_P, A_V, and A_I are the power, voltage or current gain (or loss) expressed as a ratio, P_{IN} and P_{OUT} are the input and output powers, V_{IN} and V_{OUT} are the input and output voltages, and I_{IN} and I_{OUT} are the input and output current. Note, however, that the powers, voltages or currents should be expressed in the same units/multiples (e.g. P_{IN} and P_{OUT} should *both* be expressed in W, mW, μW or nW).

It is often more convenient to express gain in decibels (rather than as a simple ratio) using the following relationships:

$$A_P = 20 \log_{10}\left(\frac{P_{OUT}}{P_{IN}}\right),\ A_V = 20 \log_{10}\left(\frac{V_{OUT}}{V_{IN}}\right) \text{and } A_I = 20 \log_{10}\left(\frac{I_{OUT}}{I_{IN}}\right)$$

Note that a *positive* result will be obtained whenever P_{OUT}, V_{OUT}, or I_{OUT} is greater than P_{IN}, V_{IN}, or I_{IN} respectively. A *negative* result will be obtained whenever P_{OUT}, V_{OUT}, or I_{OUT} is less than P_{IN}, V_{IN}, or I_{IN}. A negative result denotes *attenuation* rather than *amplification*. A *negative gain* is thus equivalent to an attenuation (or *loss*). If desired, the formulae may be adapted to produce a positive result for attenuation simply by inverting the ratios, as shown below:

$$A_P = 20 \log_{10}\left(\frac{P_{IN}}{P_{OUT}}\right),\ A_V = 20 \log_{10}\left(\frac{V_{IN}}{V_{OUT}}\right) \text{and } A_I = 20 \log_{10}\left(\frac{I_{IN}}{I_{OUT}}\right)$$

Where A_P, A_V, or A_I are the power, voltage and current gain (or loss) expressed in decibels, P_{IN} and P_{OUT} are the input and output powers, V_{IN} and V_{OUT} are the input and output voltages, and I_{IN} and I_{OUT} are the input and output current. Note, again, that the powers, voltages or currents should be expressed in the same units/multiples (e.g. P_{IN} and P_{OUT} should *both* be expressed in W, mW, μW or nW).

It is important to note that the formulae for voltage and current gain are *only* meaningful when the input and output impedances (or resistances) are identical. Voltage and current gains expressed in dB are thus *only* valid for matched constant impedance systems. The following table gives some common decibel values:

Decibels (dB)	*Power gain (ratio)*	*Voltage gain (ratio)*	*Current gain (ratio)*
0	1	1	1
1	1.26	1.12	1.12
2	1.58	1.26	1.26
3	2	1.41	1.41
4	2.51	1.58	1.58
5	3.16	1.78	1.78
6	3.98	2	2
7	5.01	2.24	2.24
8	6.31	2.51	2.51
9	7.94	2.82	2.82
10	10	3.16	3.16
13	19.95	3.98	3.98
16	39.81	6.31	6.31
20	100	10	10
30	1000	31.62	31.62
40	10000	100	100
50	100000	316.23	316.23
60	1000000	1000	1000
70	10000000	3162.3	3162.3

It should be noted that, for identical decibel values, the values of voltage and current gain can be found by taking the square root of the corresponding value of power gain.

Further values for attenuation and voltage levels in dB are given in Appendix D.

Example 3.2
A matched 600Ω attenuator produces an output of 10mV when an input of 200mV is applied. Determine the attenuation in dB.

The attenuation can be determined by applying the formula:

$A_V = 20 \log_{10}(V_{IN}/V_{OUT})$

where $V_{IN} = 200mV$ and $V_{OUT} = 10mV$.

Thus $A_V = 20 \log_{10}(200mV/10mV) = 20 \log_{10}(20) = 20 \times 1.3 = 26dB$

Example 3.3
An amplifier provides a power gain of 33dB. What output power will be produced if an input of 20mW is applied?

Here we must re-arrange the formula to make P_{OUT} the subject, as follows:

$A_P = 10 \log_{10}(P_{OUT}/P_{IN})$

thus $A_P/10 = \log_{10}(P_{OUT}/P_{IN})$

or $\text{antilog}_{10}(A_P/10) = P_{OUT}/P_{IN}$

Hence $P_{OUT} = P_{IN} \times \text{antilog}_{10}(A_P/10)$

Now $P_{IN} = 20mW = 20 \times 10^{-3}W$ and $A_P = 22dB$

$$\begin{aligned} \text{Thus } P_{OUT} &= 20 \times 10^{-3} \times \text{antilog}_{10}(22/10) \\ &= 20 \times 10^{-3} \times \text{antilog}_{10}(3.3) \\ &= 20 \times 10^{-3} \times 1.995 + 10^{3} = 39.9W \end{aligned}$$

Hints and tips

★ Fixed attenuators may be connected in cascade in order to produce an attenuation which is equal to the product of the voltage ratios or sum of the decibel values. As an example, an attenuation of 19dB can be produced by cascading 3dB, 6dB and 10dB respectively.

★ Where power levels are significant, attenuators should be constructed using resistors having an adequate power rating. The power dissipated within an attenuator will simply be the difference between the input and output power levels (P_{IN}-P_{OUT}). A 10dB attenuator supplied with an input power of 1W must, for example, be able to safely dissipate a power of 900mW.

★ In order to ensure accuracy and wide bandwidth, attenuators must employ high quality close tolerance carbon or metal film resistors. Wirewound resistors should not be used.
★ At high frequencies or where large values of attenuation are to be achieved, special care must be taken with the construction of an attenuator. Typical precautions which must be taken include screening individual sections of a multi-section attenuator, separating input and outputs as far as possible, wiring components using minimal lead lengths, employing high quality screened connectors and cables, and screening the entire attenuator assembly in a metal (grounded) enclosure.

C-R networks

Networks of capacitors and resistors (known as C-R networks) form the basis of many timing and pulse shaping circuits and are thus often encountered in practical electronic circuits.

A simple C-R circuit is shown in Figure 3.6. When the network is connected to a constant voltage source (V_s), as shown in Figure 3.7, the (initially uncharged) capacitor voltage (v_c) will rise exponentially with time, as shown in Figure 3.8. At the same time, the current in the circuit (i) will fall, as shown in Figure 3.9. The rate of charge (i.e. rate of growth of voltage with time) will be dependent upon the product of capacitance and resistance. This product is known as the *time constant* of the circuit. Hence:

time constant, $t = C \times R$

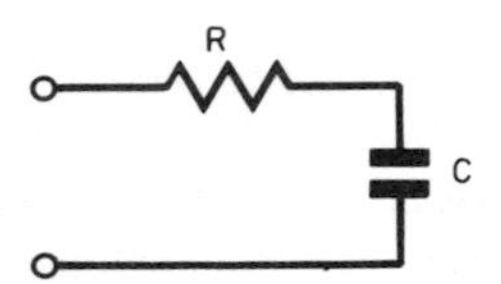

Figure 3.6 C-R circuit

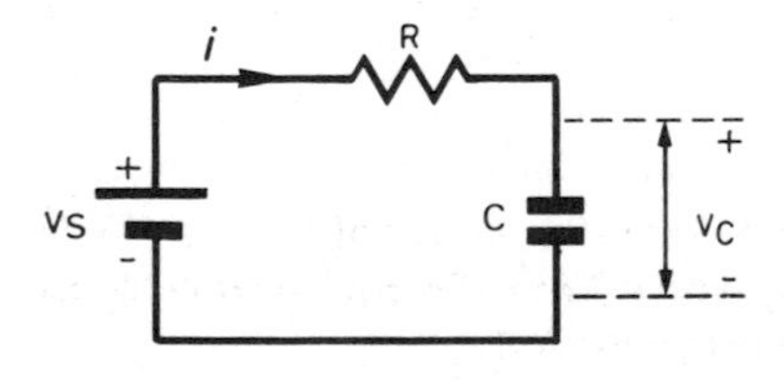

Figure 3.7 C-R circuit (C charges through R)

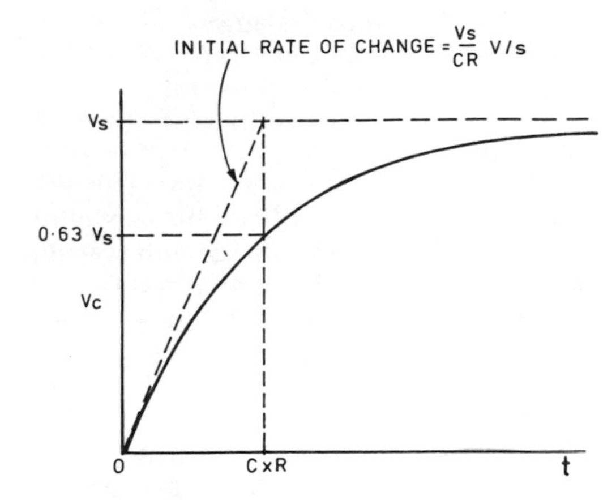

Figure 3.8 Exponential growth of capacitor voltage (v_c) in Figure 3.7

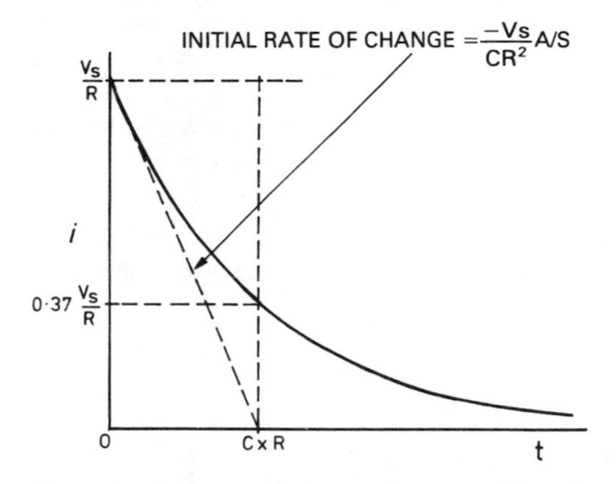

Figure 3.9 Exponential decay of current (i) in Figure 3.7

where C is the value of capacitance (in F), R is the resistance (in Ω), and t is the time constant (in s).

The voltage developed across the charging capacitor (v_c) varies with time (t) according to the relationship:

$$v_c = V_s(1 - e^{-t/CR})$$

where v_c is the capacitor voltage, V_s is the d.c. supply voltage, t is the time, and CR is the time constant of the circuit (equal to the product of capacitance (C) and resistance (R)).

The capacitor voltage will rise to approximately 63% of the supply voltage in a time interval equal to the time constant. At the end of the next interval of time equal to the time constant (i.e. after a total time of 2CR has elapsed) the voltage will have risen by a further 63% of the remainder, and so on.

In theory, the capacitor will never quite become fully charged. However, after a period of time equal to 10CR, the capacitor voltage will be very nearly equal to the supply voltage and we can consider the capacitor to be fully charged (such that $v_c \approx V_s$).

In order to simpify the mathematics of exponential growth and decay, Appendix E provides an alternative tabular method which may be used to determine the voltage and current in a C-R circuit.

A charged capacitor contains a reservoir of energy stored in the form of an electric field. When the fully charged capacitor from Figure 3.7 is connected as shown in Figure 3.10, the capacitor will discharge through the resistor, and the capacitor voltage (v_c) will fall exponentially with time, as shown in Figure 3.11. The current in the circuit (i) will also fall, as shown in Figure 3.12. The rate of

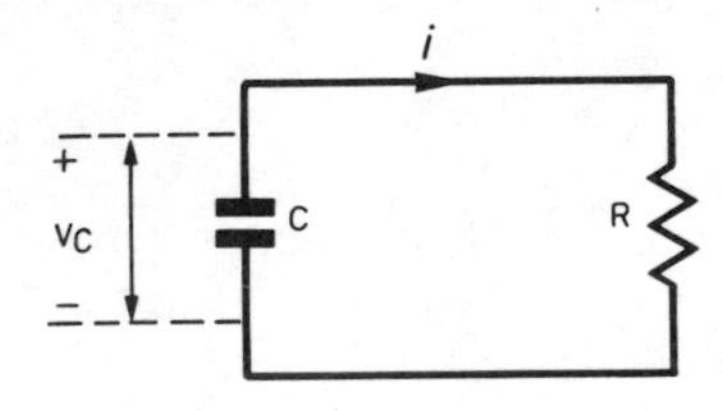

Figure 3.10 C-R circuit (C discharges through R)

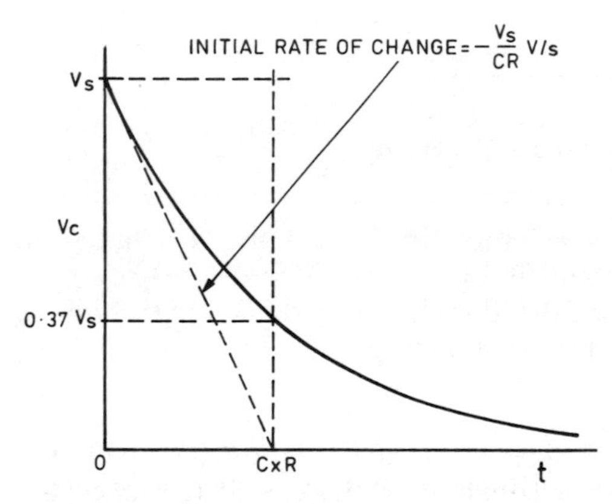

Figure 3.11 Exponential decay of capacitor voltage (V_c) in Figure 3.1

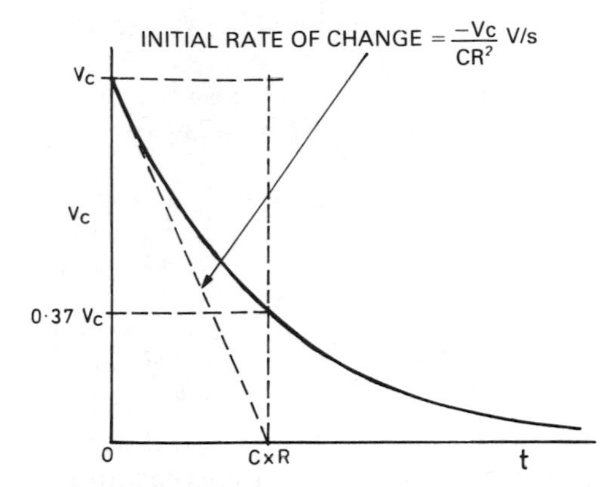

Figure 3.12 Exponential decay of current (I) in Figure 3.10

discharge (i.e. rate of decay of voltage with time) will again be governed by the time constant of the circuit (C × R).

The voltage developed across the discharging capacitor (v_c) varies with time (t) according to the relationship:

$$v_c = V e^{-t/CR}$$

where v_c is the capacitor voltage, V_s is the d.c. supply voltage, t is the time, and CR is the time constant of the circuit (equal to the product of capacitance (C) and the resistance (R)).

The capacitor voltage will fall to approximately 37% of the initial voltage in a time equal to the time constant. At the end of the next interval of time equal to the time constant (i.e. after a total time of 2CR has elapsed) the voltage will have fallen by a further 37% of the remainder, and so on.

In theory, the capacitor will never quite become fully discharged. After a period of time equal to 10CR, however, the capacitor voltage will be very nearly equal to zero and we can consider the capacitor to be fully discharged (i.e. $v_c \approx 0$).

In order to simplify the mathematics of exponential growth and decay, Appendix E provides an alternative tabular method which may be used to determine the voltage and current in a C-R circuit.

Example 3.4

An initially uncharged capacitor of 1 µF is charged from a 9V d.c. supply via a 3.3MΩ resistor. Determine the capacitor voltage 1s after connecting the supply.

The formula for exponential growth of voltage in the capacitor is:

$v_c = V_s(1 - e^{-t/CR})$

where $V_s = 9V$, $t = 1s$ and $CR = 1\mu F \times 3.3M\Omega = 3.3s$
Thus $v_c = 9(1 - e^{-1/3.3})$
or $v_c = 9(1 - 0.738)$
hence $v_c = 9 \times 0.262 = 2.358V$

Example 3.5
A 10μF capacitor is charged to a potential of 20V and then discharged through a 47kΩ resistor. Determine the time taken for the capacitor voltage to fall below 10V.

The formula for exponential decay of voltage in the capacitor is:

$c_c = V_s\ e^{-t/CR}$

where $V_s = 20V$ and $CR = 10\mu F \times 47k\Omega = 470ms$

We need to find t when $v_c = 10V$

Rearranging the formula to make t the subject gives:

$t = -CR \times \log_e(v_c/V_s)$
thus $t = -470 \times \log_e(10/20)$ ms
or $t = -470 \times -0.693 = 325ms$

Waveshaping with C-R networks

One of the most common applications of C-R networks is in simple waveshaping circuits. The circuits shown in Figures 3.13 and 3.14 function as 'square-to-triangle' and 'square-to-pulse' converters by respectively *integrating* and *differentiating* their inputs.

The effectiveness of the simple *integrator* circuit shown in Figure 3.13 depends very much upon the ratio of time constant (C×R) to periodic time (t). The *larger* this ratio is, the more effective the

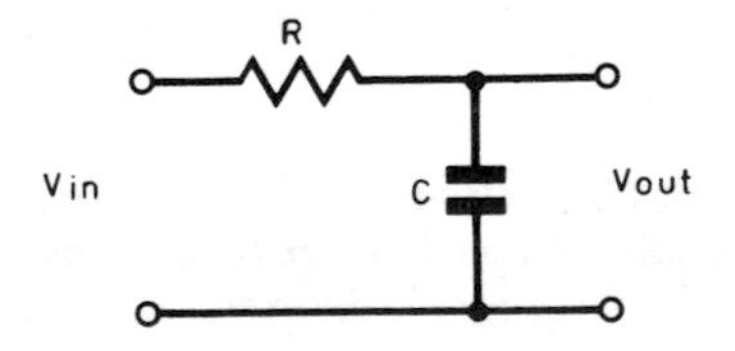

Figure 3.13 C-R- integrating circuit

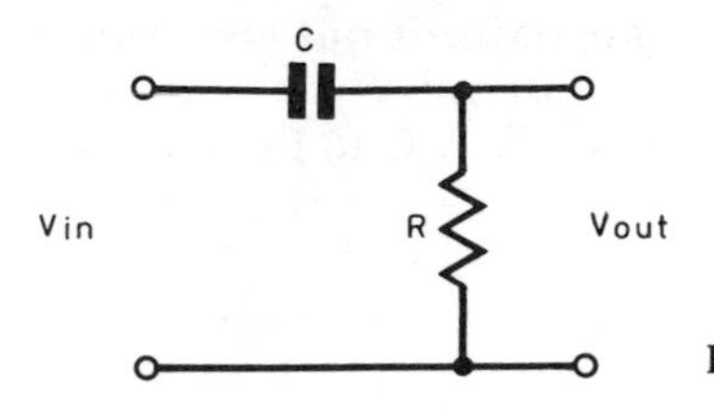

Figure 3.14 C-R differentiating circuit

Figure 3.15 Input and output waveforms for Figure 3.13

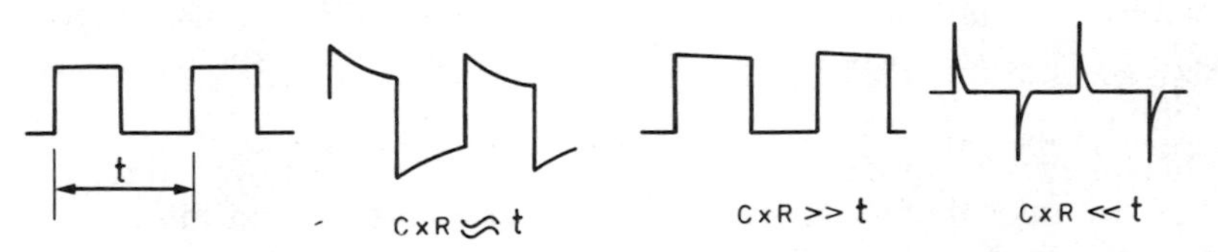

Figure 3.16 Input and output waveforms for Figure 3.14

circuit will be as an integrator. The effectiveness of the circuit of Figure 3.13 is illustrated by the input and output waveforms shown in Figure 3.15.

Similarly, the effectiveness of the simple *differentiator* circuit shown in Figure 3.14 also depends very much upon the ratio of time constant (C×R) to periodic time (t). The *smaller* this ratio is, the more effective the circuit will be as a differentiator. The effectiveness of the circuit of Figure 3.14 is illustrated by the input and output waveforms shown in Figure 3.16.

Example 3.6

A circuit is required to produce a train of alternate negative and positive pulses from a square wave of frequency 1kHz. Devise a suitable C-R arrangement and specify suitable values.

Here we require the services of a differentiating circuit along the lines of that shown in Figure 3.14. In order that the circuit operates effectively as a differentiator, we need to make the time constant

($C \times R$) very much less than the period time of the input waveform, (1ms). Assuming that we choose a medium value for R of, say 10kΩ, the maximum value which we could allow C to have would be that which satisfies the equation:

$$C \times R = 0.1\ t$$

where R = 10kΩ and t = 1ms.

$$\text{Thus } C = \frac{0.1\ t}{R} = \frac{100\mu s}{10k\Omega} = 10 \times 10^{-9}\ F = 10nF$$

In practice, any value *equal to* or *less than* 10nF would be adequate. A very small value (say below 1nF) will, however, generate pulses of a very narrow width and thus it would probably be prudent to settle for a value of either 4.7nF or 2.2nF.

Example 3.7
A circuit is required to produce a triangular waveform from a square wave of frequency 1kHz. Devise a suitable C-R arrangement and specify suitable values.

This time we require an integrating circuit like that shown in Figure 3.13. In order that the circuit operates effectively as an integrator, we need to make the time constant ($C \times R$) very much greater than the period time of the input waveform (1ms). Assuming that we choose a medium value for R of, say 10kΩ, the minimum value which we could allow C to have would be that which satisfies the equation:

$$C \times R = 10\ t$$

where R = 10kΩ and t = 1ms.

$$\text{Thus } C = \frac{10\ t}{R} = \frac{10ms}{10k\Omega} = 1 \times 10^{-6}\ F = 1\mu F$$

In practice, any value *equal to* or *greater than* 1μF would be adequate. A very large value (say above 10μF) will, however, produce a triangle wave of very severely limited amplitude thus one should settle for a value of either 2.2μF or 4.7μF.

L-R networks

Networks of inductors and resistors (known as L-R networks) can also be used for timing and pulse shaping. In comparison with capacitors, however, inductors are somewhat more difficult to manufacture and are consequently more expensive. Inductors are also prone to losses and may also require screening to minimise the effects of stray magnetic coupling. Inductors are, therefore, generally unsuited to simple timing and waveshaping applications.

Figure 3.17 shows a simple L-R network in which an inductor is connected to a constant voltage supply. When the supply is first connected, the current (i) will rise exponentially with time (as shown in Figure 3.18). At the same time, the inductor voltage (V_L) will fall (as shown in Figure 3.19). The rate of change of current with time will depend upon the ratio of inductance to resistance and is known as the *time constant*. Hence:

time constant, $t = L/R$

where L is the value of inductance (in H), R is the resistance (in Ω), and t is the time constant (in s).

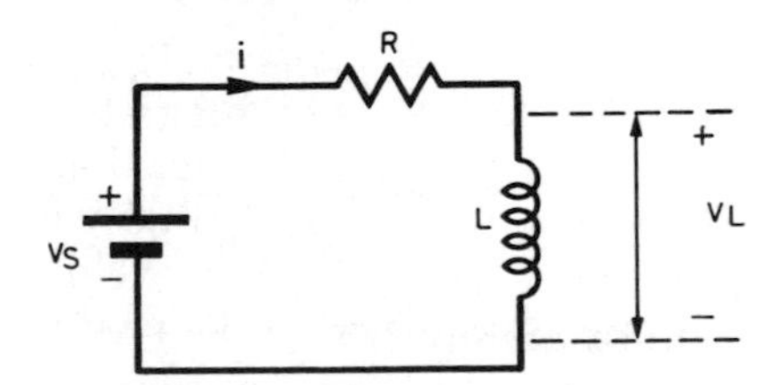

Figure 3.17 L-R circuit

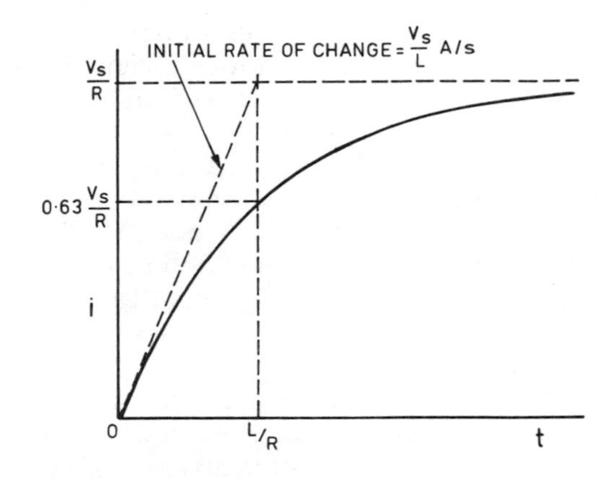

Figure 3.18 Exponential growth of current (i) in Figure 3.17

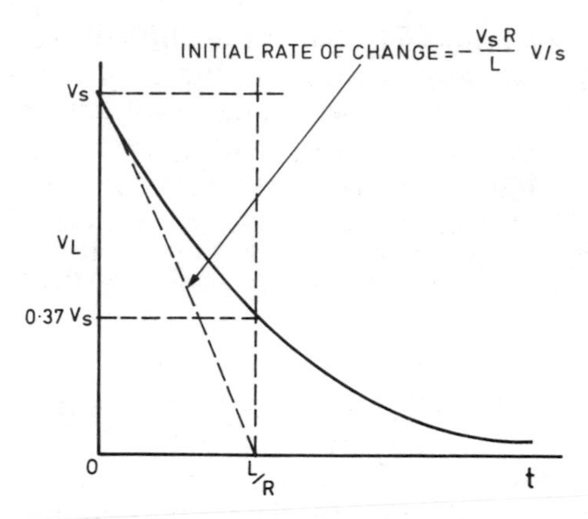

Figure 3.19 Exponential decay of inductor voltage (V_L) in Figure 3.17

The voltage developed across the inductor (v_L) varies with time (t) according to the relationship:

$$v_L = V_s \, e^{-tR/L}$$

where v_L is the inductor voltage, V_s is the d.c. supply voltage, t is the time, and L/R is the time constant of the circuit (equal to the ratio of inductance (L) to resistance (R)).

The inductor voltage will fall to approximately 37% of the initial voltage in a time equal to the time constant. At the end of the next interval of time equal to the time constant (i.e. after a total time of 2L/R has elapsed) the voltage will have fallen by a further 37% of the remainder, and so on.

The current (i) will initially be zero and will rise to approximately 63% of its maximum value (i.e. V_s/R) in a time interval equal to the time constant. At the end of the next interval of time equal to the time constant (i.e. after a total time of 2L/R has elapsed) the current will have risen by a further 63% of the remainder, and so on.

In theory, the current will never quite equal V_s/R nor will the inductor voltage ever quite fall to zero. After a period of time equal to 10CR, however, the current will be very nearly equal to V_s/R and the inductor voltage will be, to all intents and purposes, equal to zero (i.e. $v_L \approx 0$).

In order to simplify the mathematics of exponential growth and decay. Appendix E provides an alternative tabular method which may be used to determine the voltage and current in an L-R circuit.

When current is present within the inductor, energy is stored in the magnetic field. Interrupting the current (by breaking the circuit) causes the magnetic field to rapidly collapse and a high reverse voltage (known as a *back e.m.f.*) is generated momentarily across the inductor terminals. The e.m.f. generated will be directly proportional to the value of inductance and the rate of change of current with time. Hence:

$$e = L \times \frac{di}{dt}$$

where e is the induced e.m.f., L is the value of inductance, and di/dt is the rate of change of current within time. It should be noted that di/dt will be positive when the magnitude of the current is increasing and negative whenever the magnitude of the current is falling.

Example 3.8

A current of 10mA flows in an inductor of 100mH. If the current falls to zero in a time interval of 20ms, determine the e.m.f. generated across the terminals of the inductor.

The rate of change of current with time is determined as follows:

$$\frac{di}{dt} = \frac{-10\text{mA}}{20\text{ms}} = -0.5 \text{ A s}^{-1}$$

(note that di/dt is negative).

The e.m.f. generated is given by:

$$e = L \times \frac{di}{dt} = 100\text{mH} \times -0.5\text{As}^{-1} = -50\text{mV}$$

The e.m.f. generated will thus take the form of a negative-going pulse having an amplitude of 50mV.

Waveshaping with L-R networks

L-R networks are sometimes employed in waveshaping applications. The circuits shown in Figures 3.20 and 3.21 function as 'square-to-triangle' and 'square-to-pulse' converters by respectively *integrating* and *differentiating* their inputs.

The effectiveness of the simple *integrator* circuit shown in Figure 3.20 depends very much upon the ratio of time constant (L/R) to

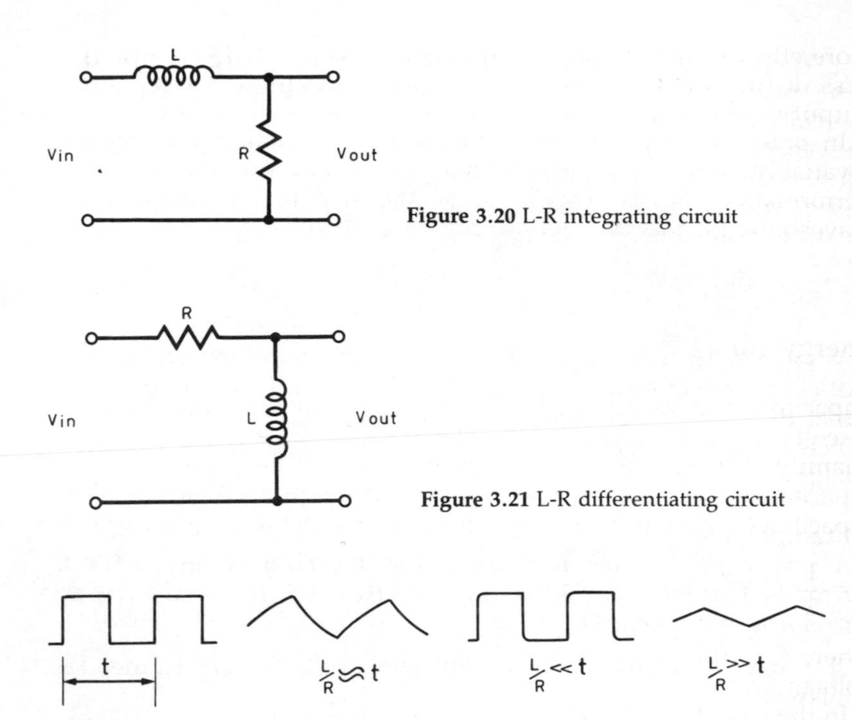

Figure 3.20 L-R integrating circuit

Figure 3.21 L-R differentiating circuit

Figure 3.22 Input and output waveforms for Figure 3.20

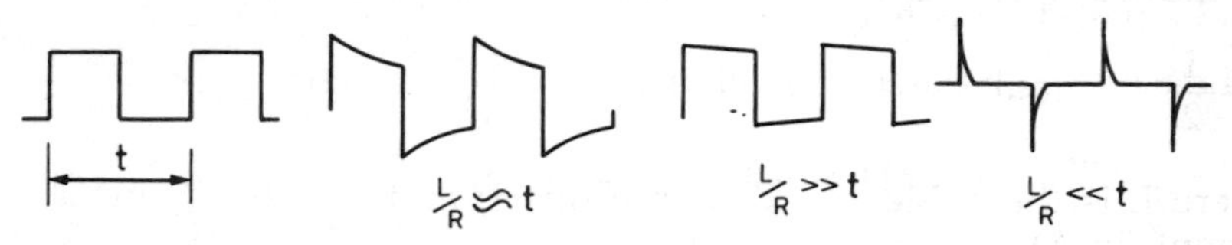

Figure 3.23 Input and output waveforms for Figure 3.21

periodic time (t). The *larger* this ratio is, the more effective the circuit will be as an integrator. The effectiveness of the circuit of Figure 3.20 is illustrated by the input and output waveforms shown in Figure 3.22.

Similarly, the effectiveness of the simple *differentiator* circuit shown in Figure 3.21 also depends very much upon the ratio of time constant (L/R) to periodic time (t). The *smaller* this ratio is, the more effective the circuit will be as a differentiator. The effectiveness of the circuit of Figure 3.21 is illustrated by the input and output waveforms shown in Figure 3.23.

In practical waveshaping applications, C-R circuits are almost invariably superior to L-R circuits on the grounds of both cost and performance. Hence examples of the use of LR circuits in waveshaping applications have not been given.

Energy storage

Capacitors and inductors are able to store electrical energy. In the case of a capacitor, the energy is stored in the electric field and the quantity of energy stored will be proportional to the product of the capacitance and the square of the voltage present across the capacitor's terminals. The amount of energy stored is given by:

$$E = \frac{1}{2} C V^2 \text{ joules}$$

where C is the value of capacitance (in F) and V is the capacitor voltage (in V).

In the case of an inductor, energy is stored in the magnetic field and the quantity of energy stored will be proportional to the product of the inductance and the square of the current flowing in the inductor. The amount of energy stored is given by:

$$E = \frac{1}{2} L I^2 \text{ joules}$$

where L is the value of inductance (in H) and I is the inductor current (in A).

A typical example of energy storage in an electronic circuit is the reservoir capacitor in a d.c. power supply. This component maintains a reservoir of charge which is continuously 'topped up' from the rectifier (usually a bridge). When the rectifier is in its non-conducting state, current is drawn from the capacitor (which effectively discharges through the load). If the supply is momentarily interrupted, the capacitor will retain its reserve of energy for several hundred milliseconds (equivalent to many tens of cycles of the mains supply). This feature helps to ensure that the operation of a circuit will continue for short periods when an a.c. supply is momentarily interrupted.

L-C networks

Two forms of L-C network are illustrated in Figure 3.24 and 3.25. Figure 3.24 is a *series resonant circuit* whilst Figure 3.25 constitutes a *parallel resonant circuit*. The impedance of both circuits varies in a complex manner with frequency.

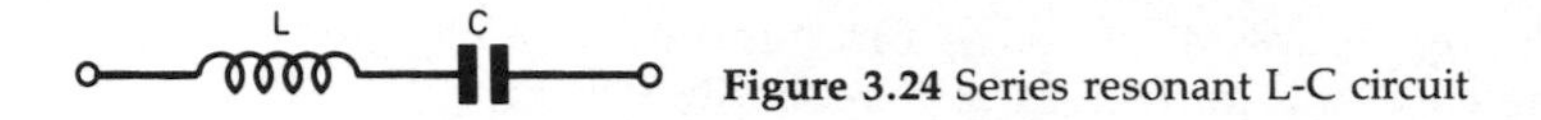

Figure 3.24 Series resonant L-C circuit

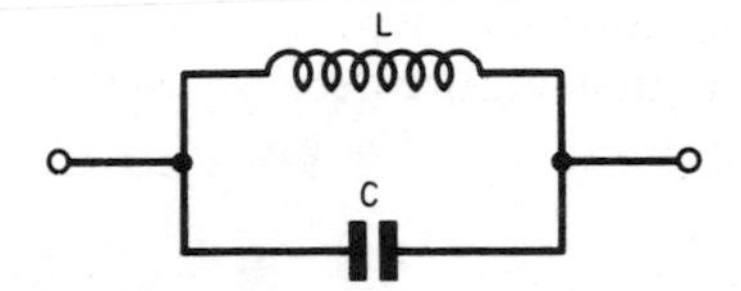

Figure 3.25 Parallel resonant L-C circuit

The impedance of the series circuit in Figure 3.24 is given by:

$$Z = \sqrt{(X_L - X_C)^2}$$

where Z is the impedance of the circuit (in Ω), and X_L and X_C are the reactance of the inductor and capacitor respectively (both expressed in Ω).

The phase angle (between the supply voltage and current) will be $+\pi/2$ rad (i.e. +90°) when $X_L > X_C$ (above resonance) or $-\pi/2$ rad (or −90°) when $X_C > X_L$ (below resonance). At a particular frequency (known as the *series resonant frequency*) the reactance of the capacitor (X_C) will be equal in magnitude (but of opposite sign) to that of the inductor (X_L). The impedance of the circuit will thus be zero at resonance. The supply current will have a maximum value at resonance (infinite in the case of a perfect parallel resonant circuit supplied from an ideal voltage source!).

The impedance of the parallel circuit in Figure 3.25 is given by:

$$Z = \frac{X_L \times X_C}{\sqrt{(X_L - X_C)^2}}$$

where Z is the impedance of the circuit (in Ω), and X_L and X_C are the reactances of the inductor and capacitor respectively (both expressed in Ω).

The phase angle (between the supply voltage and current) will be $+\pi/2$ rad (i.e. $+90°$) when $X_L > X_C$ (above resonance) or $-\pi/2$ rad (or $-90°$) when $X_C > X_L$ (below resonance).

At a particular frequency (known as the *parallel resonant frequency*) the reactance of the capacitor (X_C) will be equal in magnitude (but of opposite sign) to that of the inductor (X_L). At resonance, the denominator in the formula for impedance becomes zero and thus the circuit has an infinite impedance at resonance. The supply current will have a minimum value at resonance (zero in the case of a perfect parallel resonant circuit).

L-C-R networks

Two forms of L-C-R network are illustrated in Figures 3.26 and 3.27; Figure 3.26 is series resonant whilst Figure 3.37 is parallel

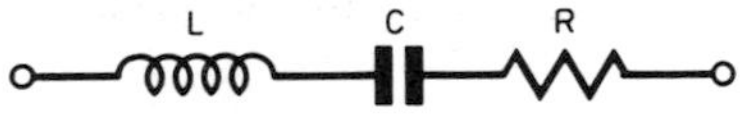

Figure 3.26 Series resonant L-C-R circuit

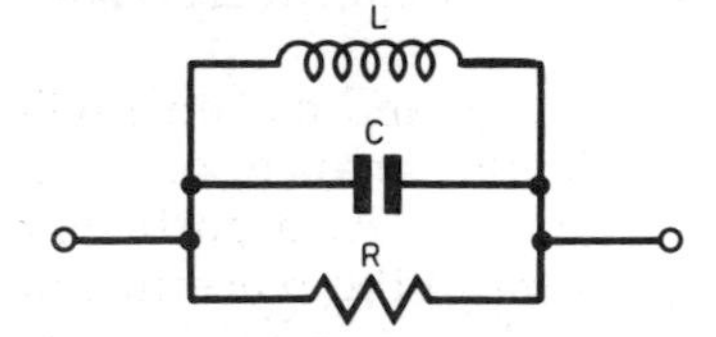

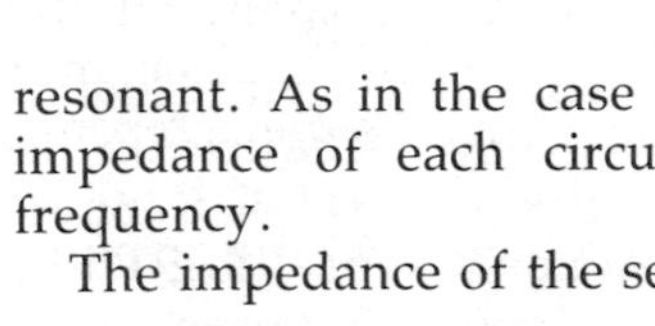

Figure 3.27 Parallel resonant L-C-R circuit

resonant. As in the case of their simpler L-C counterparts, the impedance of each circuit varies in a complex manner with frequency.

The impedance of the series circuit of Figure 3.26 is given by:

$$Z = \sqrt{R^2 + (X_L - X_C)^2}$$

where Z is the impedance of the series circuit (in Ω), R is the resistance (in Ω), X_L is the inductive reactance (in Ω) and X_C is the capacitive reactance (also in Ω). At resonance the circuit has a minimum impedance (equal to R).

The phase angle (between the supply voltage and current) will be given by:

$$\phi = \arc \tan \frac{(X_L - X_C)}{R}$$

The impedance of the parallel circuit of Figure 3.27 is given by:

$$Z = \frac{R \times X_L \times X_C}{\sqrt{(X_L{}^2 \times X_C{}^2) + R^2(X_L - X_C)^2}}$$

where Z is the impedance of the parallel circuit (in Ω), R is the resistance (in Ω), X_L is the inductive reactance (in Ω) and X_C is the capacitive reactance (also in Ω). At resonance the circuit has a maximum impedance (equal to R).

The phase angle (between the supply voltage and current) will be given by:

$$\phi = \arc \tan \frac{R\,(X_C - X_L)}{X_L \times X_C}$$

Resonance

The frequency at which the impedance is minimum for a series resonant circuit or maximum in the case of a parallel resonant circuit is known as the *resonant frequency*. The resonant frequency is given by:

$$f_o = \frac{1}{2\pi\sqrt{L\,C}}\ \text{Hz}$$

where L is the inductance (in H) and C is the capacitance (in F).

Typical impedance-frequency characteristics for series and parallel tuned circuits are shown in Figures 3.28 and 3.29. The series L-C-R tuned circuit has a minimum impedance at resonance (equal to R) and thus maximum current will flow. The circuit is consequently known as an *acceptor circuit*. The parallel L-C-R tuned circuit has a maximum impedance at resonance (equal to R) and thus minimum current will flow. The circuit is consequently known as a *rejector circuit*.

The quality of a resonant circuit (or tuned circuit) is measured by its *Q-factor*. The higher the Q-factor, the *sharper* the response (narrower bandwidth), conversely the lower the Q-factor, the *flatter* the response (wider bandwidth). In the case of the series tuned circuit,

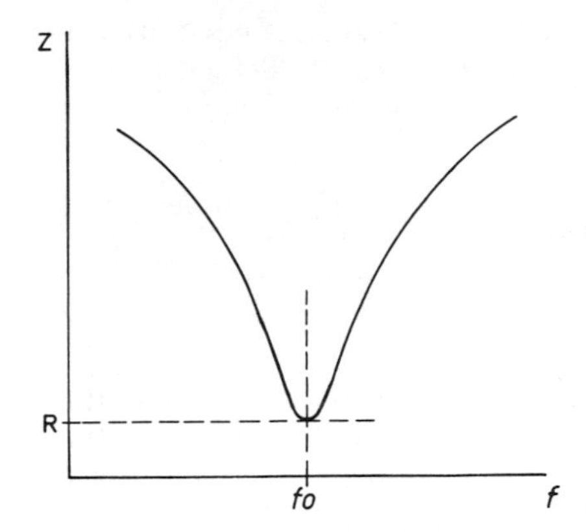

Figure 3.28 Impedance-frequency characteristic for a series resonant circuit

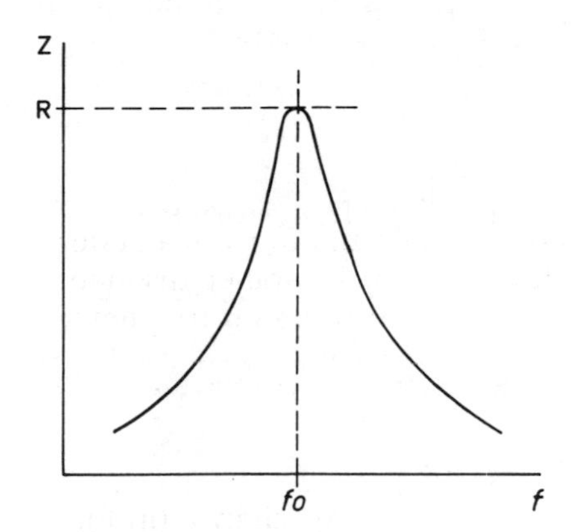

Figure 3.29 Impedance-frequency characteristic for a parallel resonant circuit

the Q-factor will increase as the resistance, R, *decreases*. In the case of the parallel tuned circuit, the Q-factor will increase as the resistance, R, *increases*. The response of a tuned circuit can be modified by incorporating a resistance of appropriate value to either 'dampen' or 'sharpen' the response.

The relationship between bandwidth and Q-factor is:

$$\text{bandwidth} = f_2 - f_1 = \frac{f_o}{Q} \text{ Hz}$$

Example 3.9
A parallel L-C circuit is to be resonant at a frequency of 400Hz. If a 100mH inductor is available, determine the value of capacitance required.

Re-arranging the formula $f_o = \frac{1}{2\pi\sqrt{LC}}$ to make C the subject gives:

$$C = \frac{1}{f_o^2 (2\pi)^2 L}$$

Thus $C = \frac{1}{400^2 \times 39.4 \times 100 \times 10^{-3}}$ F

or $C = \frac{1}{160 \times 10^3 \times 39.4 \times 100 \times 10^{-3}}$ F

Hence $C = 1.58\mu F$

this value can be realised using a 2.2μF capacitor connected in series with a 5.6μF capacitor (see Appendix C, Table 2).

Example 3.10
A series L-C-R circuit comprises an inductor of 20mH, a capacitor of 10nF, and a resistor of 100Ω. If the circuit is supplied with a sinusoidal signal of 1V at a frequency of 2kHz, determine the current supplied and the voltage developed across the resistor.

First we need to determine the values of inductive reactance (X_L) and capacitive reactance (X_C):

$$X_L = 2\pi \, f \, L = 6.28 \times 2 \times 10^3 \times 20 \times 10^{-6} \; \Omega$$

Thus $X_L = 251.2 \; \Omega$

$$X_C = \frac{1}{2\pi \, f \, C} = \frac{1}{6.28 \times 2 \times 10^3 \times 10 \times 10^{-9}} \; \Omega$$

Thus $X_C = 796.2 \; \Omega$

The impedance of the series circuit can now be calculated:

$Z = \sqrt{R^2 + (X_L - X_C)^2} = \sqrt{100^2 + (251.2 - 796.2)^2} \; \Omega$

thus $Z = \sqrt{10000 + 297025} = \sqrt{307025} = 554 \; \Omega$

The current flowing in the series circuit is given by:

$i = V/Z = 1V/554\Omega = 1.8mA$

The voltage developed across the resistor can now be calculated using:

$V = I\,R = 1.8\text{mA} \times 100\Omega = 180\text{mV}$

Filter types and characteristics

Filters provide a means of restricting the bandwidth of a signal or selecting signals within a particular frequency range. Various types of filter are commonly encountered and the frequency response of each is illustrated in Figure 3.30. Figure 3.30(a) shows the characteristics of ideal filters whilst Figure 3.30(b) shows the characteristics of real filters. Figure 3.31 shows the conventional block symbols for each of the filters shown in Figure 3.30.

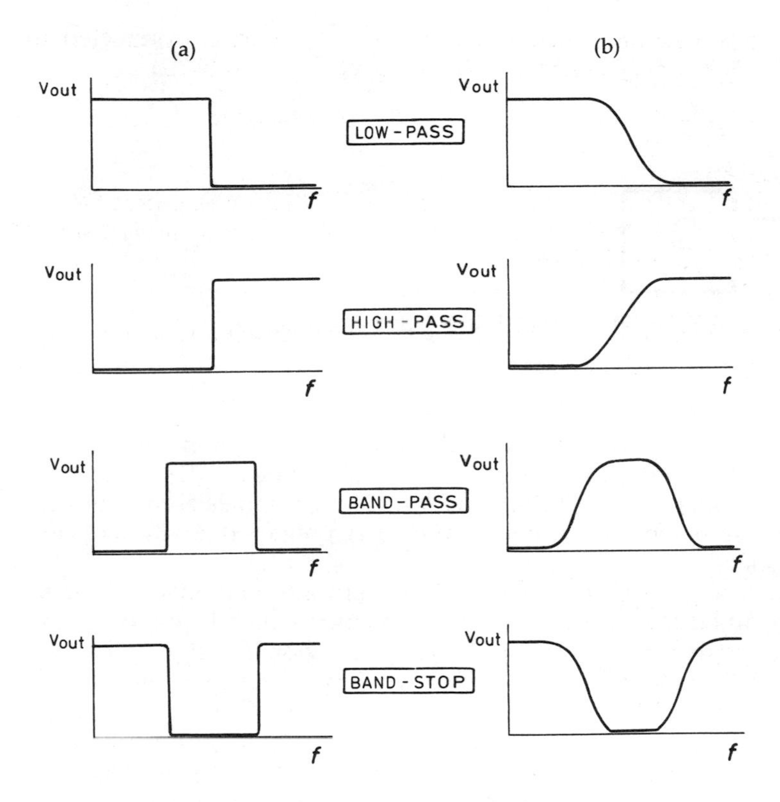

Figure 3.30 Filter characteristics: (a) ideal; (b) real

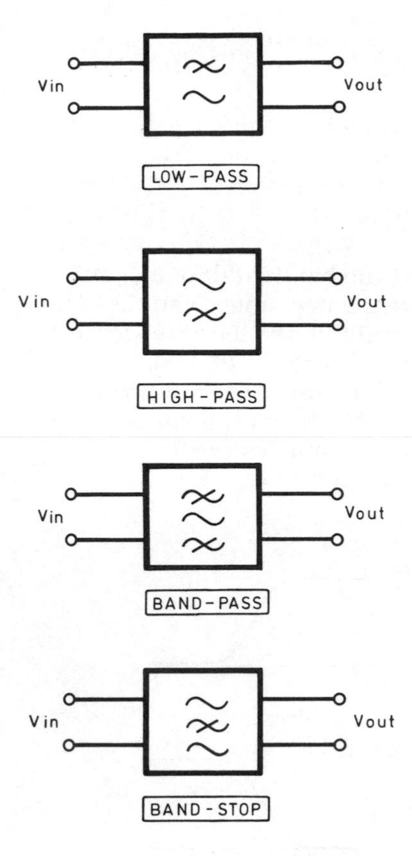

Figure 3.31 Filter symbols

An ideal *low-pass filter* will exhibit zero attenuation for signals below cut-off (i.e. those having a frequency which is less than the cut-off frequency) and infinite attenuation for signals above cut-off (i.e. those having a frequency which is greater than the cut-off frequency).

An ideal *high-pass filter* will exhibit zero attenuation for signals above cut-off (i.e. those having a frequency which is greater than the cut-off frequency) and infinite attenuation for signals below cut-off (i.e. those having a frequency which is less than the cut-off frequency).

An ideal *band-pass filter* will exhibit zero attenuation for signals between the two cut-off frequencies and infinite attenuation beyond these values.

An ideal *band-stop filter* will exhibit infinite attenuation for signals between the two cut-off frequencies and zero attenuation outside these limits.

The *cut-off frequency* of a filter is defined as the frequency at which the output power of the filter falls to 50% (or −3dB) of its input value. The cut-off frequency is thus sometimes also referred to as the *half-power frequency* or *−3dB frequency*.

Low-pass and high-pass filters have just one cut-off frequency whilst band-pass types will have a lower cut-off frequency (f_1) and an upper cut-off frequency (f_2). The rate at which the response of a filter falls beyond its cut-off point is important in determining the effectiveness of the filter. Ideally, the slope of the filter's response curve should be infinite (i.e. the filter should exhibit no response beyond the cut-off point). In practice the slope of the filter response tends to fall at a rate of −6dB per octave for a passive first-order filter, and at −12dB (or more) per octave for an active second-order filter.

The *bandwidth* of a band-pass or band-stop filter is equal to the difference between the cut-off frequencies (i.e. $f_2 - f_1$).

Prototype filters

Figures 3.32 to 3.35 show how each of the filter types in Figure 3.30 can be realised using simple networks of capacitors and

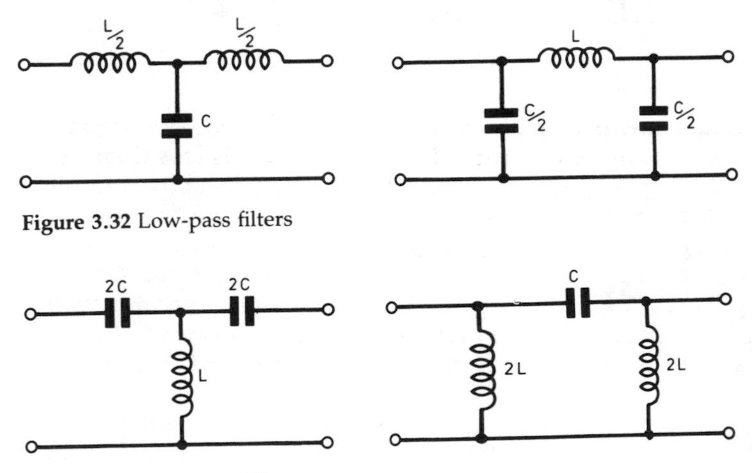

Figure 3.32 Low-pass filters

Figure 3.33 High-pass filters

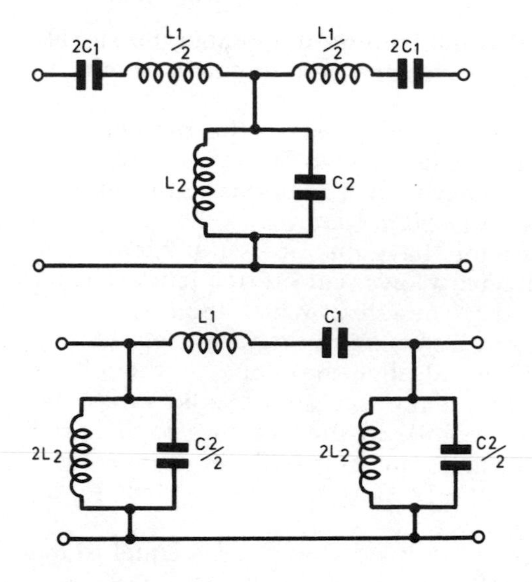

Figure 3.34 Band-pass filters

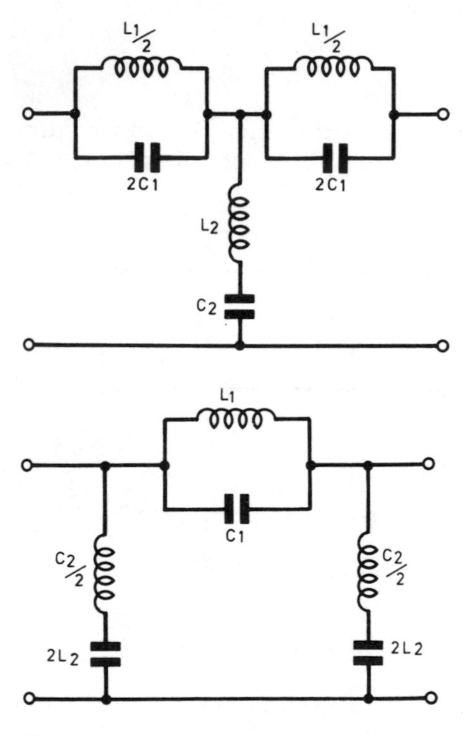

Figure 3.35 Band-stop filters

inductors in either T-section or π-section configurations. The formulae for these prototype constant-k filters are as follows:

Low-pass (Figure 3.32)

$$L = \frac{Z_o}{\pi \, f_c} \text{ H}$$

$$C = \frac{1}{\pi \, Z_o \, f_c} \text{ F}$$

$$f_C = \frac{1}{\pi \sqrt{L\,C}} \text{ Hz}$$

High-pass (Figure 3.33)

$$L = \frac{Z_o}{4 \, \pi \, f_C} \text{ H}$$

$$C = \frac{1}{4 \, \pi \, Z_o \, f_C} \text{ F}$$

$$f_C = \frac{1}{4 \, \pi \sqrt{L\,C}} \text{ Hz}$$

Band-pass (Figure 3.34)

$$L_1 = \frac{Z_o}{\pi \, (f_2 - f_1)} \text{ H}$$

$$L_2 = \frac{Z_o \, (f_2 - f_1)}{4 \, \pi \, f_1 \, f_2} \text{ H}$$

$$C_1 = \frac{(f_2 - f_1)}{4 \, \pi \, f_1 \, f_2 \, Z_o} \text{ F}$$

$$C_2 = \frac{1}{\pi \, Z_o \, (f_2 - f_2)} \text{ F}$$

Band-stop (Figure 3.35)

$$L_1 = \frac{Z_o \, (f_2 - f_1)}{\pi \, f_1 \, f_2} \text{ H}$$

$$L_2 = \frac{Z_o}{4 \, \pi \, (f_2 - f_1)} \text{ H}$$

$$C_1 = \frac{1}{4 \pi Z_o (f_2 - f_1)} \text{ F}$$

$$C_2 = \frac{(f_2 - f_1)}{\pi f_1 f_2 Z_o} \text{ F}$$

Where f_C is the cut-off frequency (in the case of low-pass and high-pass filters), and f_1 and f_2 are the lower and upper cut-off frequencies respectively (in the case of band-pass and band-stop filters).

The *design impedance* of each of the above filters is given by:

$$Z_o = \sqrt{\frac{L}{C}}$$

Example 3.11

A π-section low-pass filter is to be constructed according to the following specification:

Design impedance: 600Ω
Cut-off frequency: 1kHz

If an inductor of 60mH is available, determine the value of the capacitors required.

Using the π-section prototype constant-k filter shown in Figure 3.32 and its associated design formulae gives:

$$C = \frac{1}{\pi Z_o f_c} = \frac{1}{3.142 \times 600 \times 1 \times 10^3} \text{ F}$$

Hence $C = 5.3 \times 10^{-7}$ F = 0.53μF

In the π-section configuration, each individual capacitor must have a value equal to 2C. Hence the required values will be 1.06μF (in practice, 1μF capacitors would be perfectly adequate).

Hints and tips

★ Filter sections may be cascaded to provide greater slopes. As an example, two passive filters (each providing −6dB per octave) may be connected in tandem to provide a filter which will have a nominal −12dB per octave response beyond cut-off.

★ In order to ensure accuracy filters must employ high quality close tolerance inductors and capacitors.

★ In order to maintain filter performance at frequencies well beyond cut-off (i.e. to ensure ultimate values of attenuation are achieved) special care must be taken with the layout and construction of an attenuator. Typical precautions which must be taken include screening individual sections of a multi-section attenuator (to avoid unwanted inductive or capacitive coupling), separating input and outputs as far as possible, wiring components using minimal lead lengths, employing high quality screened connectors and cables, and screening the entire filter assembly in a metal (grounded) enclosure.

4 Diodes

Diode characteristics

Semiconductor diodes generally comprise a single p-n junction of either silicon (Si) or germanium (Ge) material. In order to obtain conduction, the p-type material must be made positive with respect to the n-type material (the p-type connection constitutes the anode whilst the n-type connection constitutes the cathode). The direction of current flow is from anode to cathode when the diode is conducting, as shown in Figure 4.1. Very little current (negligible in the case of most silicon devices) flows in the reverse direction (Figure 4.2).

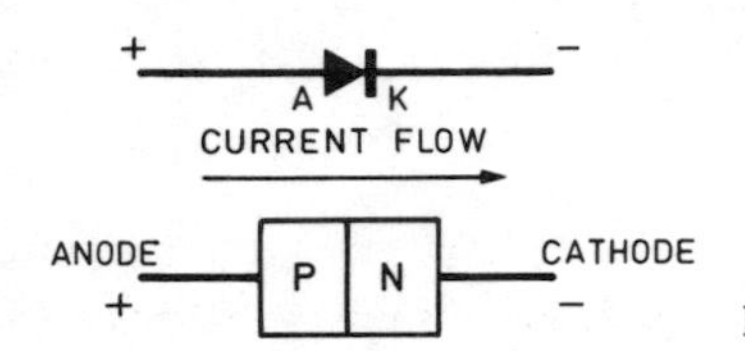

Figure 4.1 Forward biased (conducting) diode

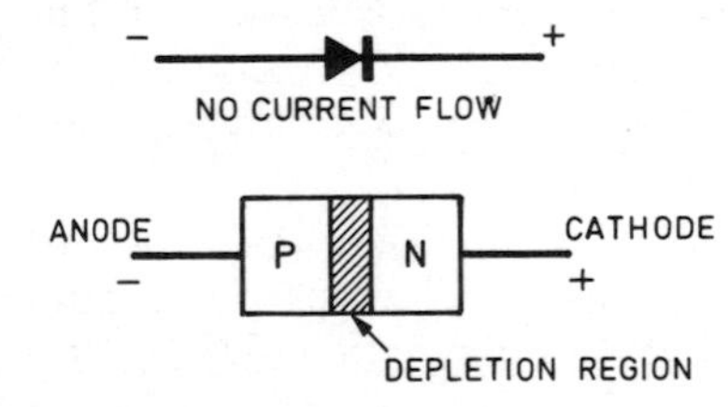

Figure 4.2 Reverse biased (non-conducting) diode

Diodes exhibit a low resistance to current flow in one direction and a high resistance to current flow in the other. The direction in which current flows is referred to as the *forward* direction whilst that in which negligible current flows is known as the *reverse* direction. When a diode is conducting, a diode is said to be forward biased and a small voltage (ideally zero) is dropped across it. This voltage is known as the forward voltage drop. The maximum reverse voltage that a diode can tolerate is usually specified in terms of its reverse repetitive maximum voltage (V_{RRM}) or peak inverse voltage (PIV).

Typical values of forward current and forward voltage for commonly available silicon and germanium diodes are given below:

Forward current	*Forward voltage drop*		
	Silicon (1N4148)	*Silicon (1N5401)*	*Germanium (OA91)*
10µA	0.43V	–	0.12V
100µA	0.58V	0.55V	0.26V
1mA	0.65V	0.60V	0.32V
10mA	0.75V	0.65V	0.43V
100mA	–	0.72V	–
1A	–	0.85V	–

Germanium diodes conduct at lower forward voltages than their silicon counterparts (typically 100mV as compared with 600mV), but they tend to exhibit considerably more reverse leakage current (1µA as compared with 10nA for an applied reverse voltage of 50V). Furthermore, the forward resistance of a conducting silicon diode is much lower than that of a comparable germanium type. Hence germanium diodes are used primarily for signal detection purposes whereas silicon devices are used for rectification and for general purpose applications. Typical forward and reverse characteristics for comparable germanium and silicon diodes are shown in Figure 4.3.

Diodes are often divided into signal and rectifier types, according to their principal field of application. *Signal diodes* require consistent forward characteristics with low forward voltage drop.

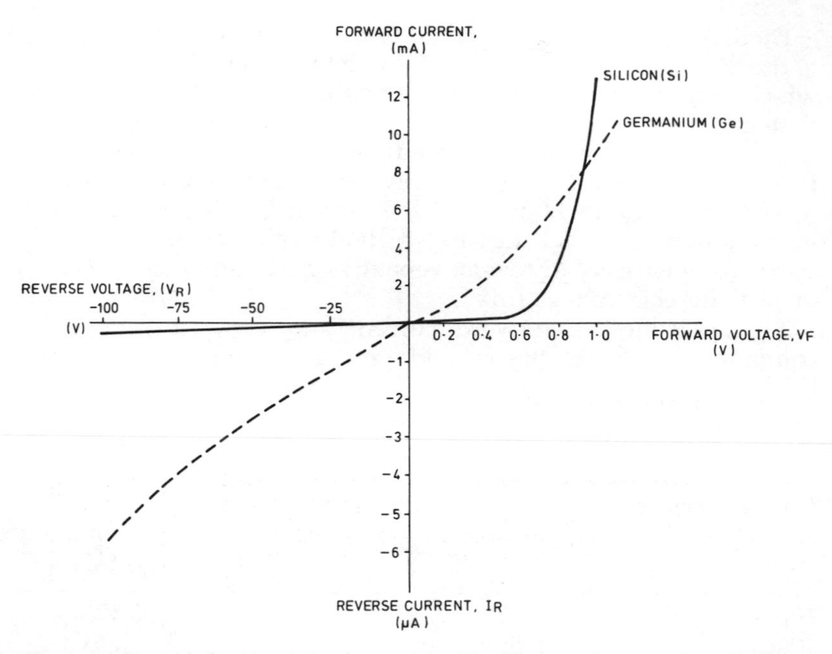

Figure 4.3 Typical characteristics for comparable silicon and germanium diodes

Figure 4.4 Bridge rectifier arrangement

Rectifier diodes need to be able to cope with high values of reverse voltage and large values of forward current, consistency of characteristics is of secondary importance in such applications. Rectifier diodes are often available in the form of a *bridge* (see Figure 4.4) which provides full-wave rectification. Various diode encapsulations are illustrated in Figure 4.5.

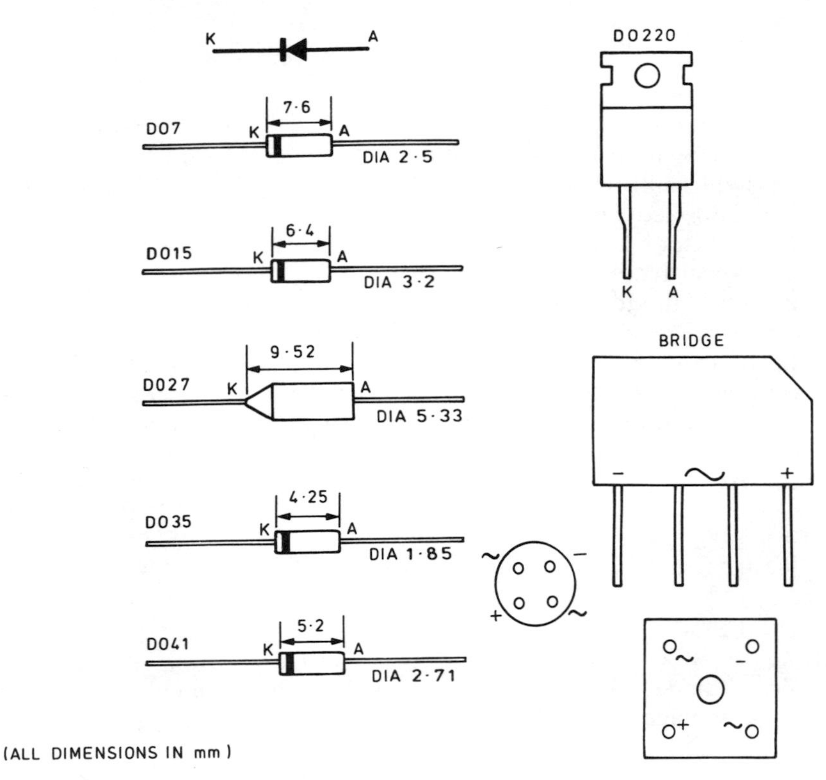

Figure 4.5 Diode encapsulations

Diode coding

The European system for classifying semiconductor diodes involves an alphanumeric code which employs either two letters and three figures (general purpose diodes) or three letters and two figures (special purpose diodes). The first two letters have the following significance:

First letter – semiconductor material:

A germanium
B silicon
C gallium arsenide etc
D photodiodes etc

Second letter – application:
A general purpose diode
B tuning (varicap) diode
E tunnel diode
P photovoltaic diode
Q light emitting diode
T controlled rectifier
X varactor diode
Y power rectifier
Z zener diode

In the case of diodes for specialised applications, the third letter does not generally have any particular significance. Zener diodes have an additional letter (which appears *after* the numbers) which denotes the tolerance of the zener voltage. The following letters are used:

A ±1%
B ±2%
C ±5%
D ±10%

Zener diodes also have additional characters which indicate the zener voltage (e.g. 9V1 denotes 9.1V).

Example 4.1
Identify each of the following diodes:
(i) AA113
(ii) BB105
(iii) BZY88C4V7

Diode (i) is a general purpose germanium diode.
Diode (ii) is a silicon diode for tuning applications (sometimes referred to as a varicap).
Diode (iii) is a silicon zener diode having 5% tolerance and 4.7V zener voltage.

Diode data

The following tables summarise the characteristics of a variety of popular semiconductor diodes:

Devices, applications, equivalents and case styles

Device	*Material*	*Application*	*Near equiv.*	*Case style*
1N4001	silicon	rectifier		DO41
1N4002	silicon	rectifier		DO41
1N4003	silicon	rectifier		DO41
1N4004	silicon	rectifier		DO41
1N4005	silicon	rectifier		DO41
1N4006	silicon	rectifier		DO41
1N4007	silicon	rectifier	BY127	DO41
1N4148	silicon	general purpose	1N914, 1N916	DO35
1N5400	silicon	rectifier		DO27
1N5401	silicon	rectifier		DO27
1N5402	silicon	rectifier		DO27
1N5404	silicon	rectifier		DO27
1N5406	silicon	rectifier		DO27
1N5407	silicon	rectifier		DO27
1N5408	silicon	rectifier		DO27
1N914	silicon	general purpose	1N916, 1N4148	DO35
1N916	silicon	general purpose	1N914, 1N4148	DO35
AA113	germanium	general purpose		DO7
AA119	germanium	RF detector		DO7
BAR28	Schottky	RF detector		DO35
BAX13	silicon	general purpose		DO35
BAX16	silicon	general purpose		DO35
BY126	silicon	rectifier	1N4005	DO15
BY127	silicon	rectifier	1N4007	DO15
HSCH1001	Schottky	RF detector		DO35
OA200	silicon	general purpose		DO7
OA202	silicon	general purpose		DO7
OA47	germanium	general purpose		DO7
OA90	germanium	RF detector		DO7
OA91	germanium	general purpose	OA95	DO7
OA95	germanium	general purpose	OA91	DO7

General purpose, signal and RF diodes

Device	*Material*	*PIV*	I_f	I_r *max*
1N4148	silicon	100V	75mA	25nA
1N914	silicon	100V	75mA	25nA
1N916	silicon	100V	75mA	25nA
AA113	germanium	60V	10mA	200μA
AA119	germanium	45V	35mA	350μA
BAR28	Schottky	70V		200nA
BAX13	silicon	50V	75mA	200nA

continued overleaf

Device	*Material*	*PIV*	I_f	I_r *max*
Bax16	silicon	150V	200mA	100nA
HSCH1001	Schottky	60V	15mA	200nA
OA200	silicon	50V	80mA	100nA
OA202	silicon	150V	40mA	100nA
OA47	germanium	25V	110mA	100μA
OA90	germanium	30V	10mA	1.1mA
OA91	germanium	115V	50mA	275μA
OA95	germanium	115V	50mA	250μA

Silicon rectifier and power diodes

Device	*PIV*	I_{fav}	V_f	I_r *max*
1N4001	50V	1A	1.1V	10μA
1N4002	100V	1A	1.1V	10μA
1N4003	200V	1A	1.1V	10μA
1N4004	400V	1A	1.1V	10μA
1N4005	600V	1A	1.1V	10μA
1N4006	800V	1A	1.1V	10μA
1N4007	1000V	1A	1.1V	10μA
1N5400	50V	3A	1.1V	10μA
1N5401	100V	3A	1.1V	10μA
1N5402	200V	3A	1.1V	10μA
1N5404	400V	3A	1.1V	10μA
1N5406	600V	3A	1.1V	10μA
1N5407	800V	3A	1.1V	10μA
1N5408	1000V	3A	1.1V	10μA
BY126	650V	1A	1.1V	10μA
BY127	1250V	1A	1.1V	10μA
BY397	200V	2A	1.1V	10μA
BY399	800V	2A	1.1V	10μA

Bridge rectifiers

Type/series	*Encapsulation*	*Mounting surface*	*Max. forward current (A)*
Vm	4-pin d.i.l.	PCB	0.9
DB	4-pin d.i.l.	PCB	1
WO	cylindrical	PCB	1
SKB2	in-line	PCB	1.6
BR8	in-line	PCB	2
BR3	square	PCB	3

Type/series	*Encapsulation*	*Mounting surface*	*Max. forward current (A)*
KBPC	square	PCB	2 to 6
BR6	square	PCB	6
BR15	epoxy-potted	heatsink	15
SKB25	epoxy-potted	heatsink	6 to 35

Note: Most of the bridge rectifiers listed are available in 200V, 400V and 600V versions. It is important to ensure that manufacturers' voltage ratings are not exceeded.

Hints and tips

★ When designing power supply circuits (in which appreciable currents are present) it is important to allow for the forward voltage drop associated with each rectifier diode. In a bridge rectifier, for example, two diodes will be conducting at any one time. The total forward voltage drop associated with these diodes can approach 2V and this should be allowed for when determining the a.c. input voltage to the rectifier.

★ The reverse leakage current of a diode increases markedly as the junction temperature increases. This results in a reduction in overall efficiency (ratio of forward current to reverse current) at high temperatures.

★ Operating a diode at, or beyond, the stated limits for V_{RRM} or PIV will result in a high risk of breakdown. Since rectifier failure can have disastrous consequences, it is always advisable to operate diodes well within the stated limits (to ensure safety, a 100% margin should be allowed).

★ Schottky diodes exhibit a forward voltage drop which is approximately half that of a conventional silicon diodes coupled with very fast reverse recovery. Schottky diodes are thus preferred in switching applications (e.g. switched mode power supplies) where very low forward voltage drop and fast switching is a prime consideration.

Zener diodes

Zener diodes are silicon diodes which are specially designed to exhibit consistent *reverse breakdown* characteristics. Zener diodes are available in various families (according to their general characteristics, encapsulation and power ratings) with reverse breakdown (zener) voltages in the E12 and E24 series (ranging

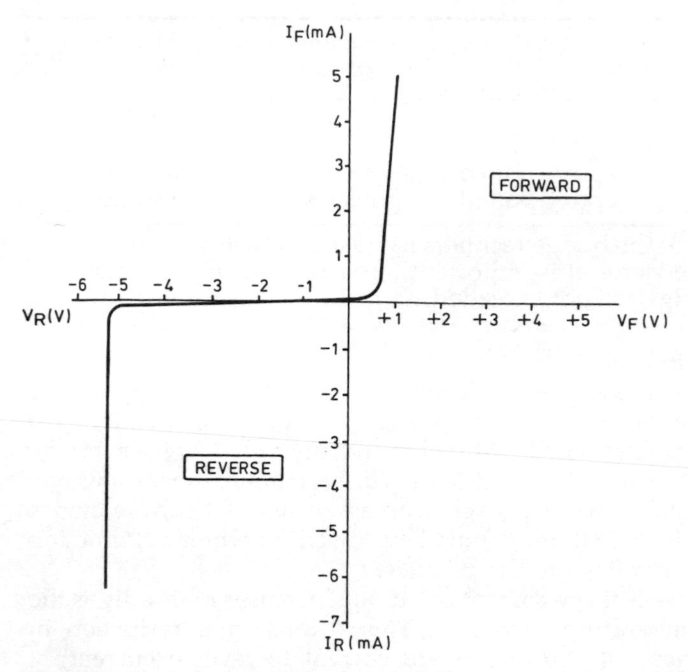

Figure 4.6 Typical zener diode characteristics

from 2.4V to 91V). A typical characteristic for a 5.1V zener diode is shown in Figure 4.6.

The following series of zener diodes are commonly available:

BZY88 series
Miniature glass encpsulated diodes rated at 500mW (at 25 deg.C). Zener voltages range from 2.7V to 15V (voltages are quoted for 5mA reverse current at 25 deg.C).

BZX55 series
Low-power diodes rated at 500mW and offering zener voltages in the range 2.4V to 91V.

BZX61 series
Encapsulated alloy junction rated at 1.3W (25 deg.C ambient). Zener voltages range from 7.5V to 72V.

BZX85 series
Medium-power glass-encapsulated diodes rated at 1.3W and offering zener voltages in the range 5.1V to 62V.

BZY93 series
High power diodes in stud mounting encapsulation. Rated at 20W for ambient temperatures up to 75 deg.C. Zener voltages range from 9.1V to 75V.

BZY97 series
Medium power wire-ended diodes rated at 1.5W and offering zener voltages in the range 9.1V to 37V.

1N5333 series
Plastic encapsulated diodes rated at 5W. Zener voltages range from 3.3V to 24V.

Zener diodes are generally plastic or glass encapsulated in the same manner as conventional silicon diodes. As with conventional silicon diodes, the cathode connection is marked with a stripe (see Figure 4.7).

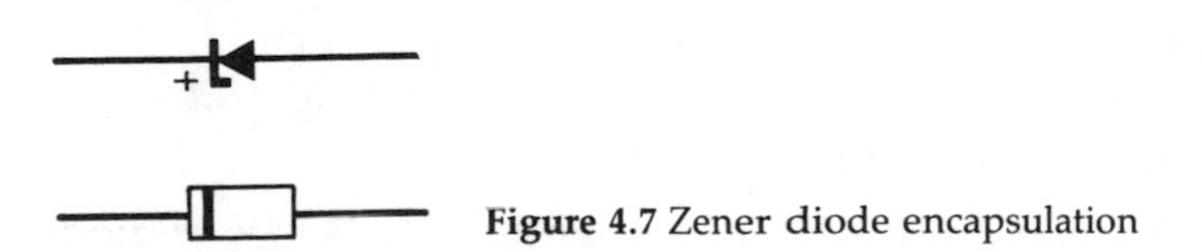

Figure 4.7 Zener diode encapsulation

The *slope resistance* of a zener diode is the rate of change of reverse voltage (zener voltage) with diode current. Slope resistance is measured in the breakdown region and expressed in ohms. An ideal zener diode would have zero slope resistance (i.e. the diode would conduct perfectly at its rated zener voltage). In practice, values of 20Ω, or less, can be achieved.

The *temperature coefficient* of zener voltage is the change of zener voltage (from its rated value) which results from a temperature change of 1°C. Temperature coefficient (which should ideally be zero) is expressed in mV/°C. In many voltage reference applications, it is essential for the reference diode to exhibit a zener voltage which does not vary with temperature. The following data (for the BZX55 series) is typical of most low-power zener diodes:

Zener voltage (V)	*Slope resistance (Ω)*	*Temperature coefficient (mV/°C)*
2.7	100	−3.5
3.3	95	−3.5
3.9	90	−3.5
4.7	80	−3.5
5.1	60	−2.7
5.6	40	−2.0
6.2	10	+0.4
6.8	12	+1.2
7.5	14	+2.5
8.2	16	+3.2
9.1	18	+3.8
10	20	+4.5

Example 4.2

A zener diode is to be used as a voltage reference. The diode has the following specifications:

Zener voltage (at 20°C): *9.1V*
Temperature coefficient: *+4mV/°C*

If the equipment is designed to operate over the range −10°C to +40°C, determine the extreme values of reference voltage and the percentage change in reference voltage over the working range.

The temperature coefficient is positive and thus the zener voltage will increase with temperature. At 40°C the zener voltage will be given by:

$$V_z = 9.1V + ((40-20) \times 4mV)$$
$$= 9.1V + 80mV = 9.18V$$

At −10°C the zener voltage will be given by:

$$V_z = 9.1V - ((20--10) \times 4mV)$$
$$= 9.1V - 120mV = 8.98V$$

The total change in temperature will be 50°C and the corresponding change in zener voltage will be 50 × 4mV or 200mV. The percentage change will thus be given by:

$$\% \text{ change} = \frac{200mV}{9.1V} \times 100 = 2.2\%$$

Hints and tips

★ Zener diodes may be connected in series to obtain higher voltages. As an example, a 15.9V reference can be produced by connecting a 6.8V zener diode in series with a 9.1V zener diode.

★ Care must be taken to ensure that zener diodes operate within their rated power dissipation (see Section 7 for further details).

★ Zener diodes generally perform best when rated at voltages of between 5V and 6V. Hence, in order to obtain optimum performance (in terms of both slope resistance and temperature coefficient) reference voltage sources based upon zener diodes should utilise components which have zener voltages of between 5.1V and 6.2V. Where necessary, external circuitry can be used to provide voltage amplification.

★ Zener diodes can generate a significant amount of noise and, in applications which involve significant voltage gain (e.g. the stabilisation of an amplifier bias supply) it is essential to provide adequate decoupling. A parallel connected capacitor of between 1μF and 100μF will prove effective in most applications.

Thyristors

Thyristors (or silicon controlled rectifiers) are three-terminal devices which can be used for switching and a.c. power control. Thyristors can switch very rapidly from a non-conducting to a conducting state. In the off state, the thyristor exhibits negligible leakage current whilst, in the on state the device exhibits very low resistance. This results in very little power loss within the thyristor even when appreciable power levels are being controlled. Once switched into the conducting state, the thyristor will remain conducting (i.e. it is latched in the on state) until the forward current is removed from the device. In d.c. applications this

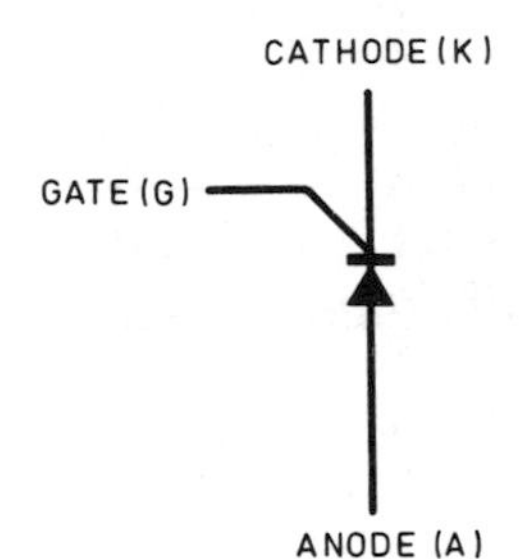

Figure 4.8 Thyristor connections

necessitates the interruption (or disconnection) of the supply before the device can be reset into its non-conducting state. Where the device is used with an alternating supply, the device will automatically become reset whenever the main supply reverses. The device can then be triggered on the next half-cycle having correct polarity to permit conduction. Like their conventional silicon diode counterparts, thyristors have anode and cathode connections; control is applied by means of a *gate* terminal (see Figure 4.8). The device is *triggered* into the conducting (on state) by means of the application of a current pulse to this terminal.

Thyristor data

The table summarises the characteristics of a variety of popular thyristors:

Type	$I_{F(AV)}$	V_{RRM}	V_{GT}	I_{GT}	*Case style*
2N4443	5.1A	400V	1.5V	30mA	TO220
2N4444	5.1A	600V	1.5V	30mA	TO220
BT106	1A	700V	3.5V	50mA	Stud
BT152	13A	600V	1V	32mA	TO220
BTX18-400	1A	500V	2V	5mA	TO5
BTY79-400R	6.4A	400V	3V	30mA	Stud
BTY79-600R	6.4A	600V	3V	30mA	Stud
BTY79-800R	6.4A	800V	3V	30mA	Stud
TIC106A	3.2A	100V	1.2V	200μA	TO220
TIC106B	3.2A	200V	1.2V	200μA	TO220
TIC106C	3.2A	300V	1.2V	200μA	TO220
TIC106D	3.2A	400V	1.2V	200μA	TO220
TIC106E	3.2A	500V	1.2V	200μA	TO220
TIC106M	3.2A	600V	1.2V	200μA	TO220
TIC106S	3.2A	700V	1.2V	200μA	TO220
TIC106N	3.2A	800V	1.2V	200μA	TO220
TIC116A	5A	100V	2.5V	20mA	TO220
TIC116B	5A	200V	2.5V	20mA	TO220
TIC116C	5A	300V	2.5V	20mA	TO220
TIC116D	5A	400V	2.5V	20mA	TO220
TIC116E	5A	500V	2.5V	20mA	TO220
TIC116M	5A	600V	2.5V	20mA	TO220
TIC116S	5A	700V	2.5V	20mA	TO220
TIC116N	5A	800V	2.5V	20mA	TO220

Type	$I_{F(AV)}$	V_{RRM}	V_{GT}	I_{GT}	*Case style*
TIC126A	7.5A	100V	2.5V	20mA	TO220
TIC126B	7.5A	200V	2.5V	20mA	TO220
TIC126C	7.5A	300V	2.5V	20mA	TO220
TIC126D	7.5A	400V	2.5V	20mA	TO220
TIC126E	7.5A	500V	2.5V	20mA	TO220
TIC126M	7.5A	600V	2.5V	20mA	TO220
TIC126S	7.5A	700V	2.5V	20mA	TO220
TIC126N	7.5A	800V	2.5V	20mA	TO220
TICP106D	2A	400V	1V	200μA	TO92
TICP106M	2A	600V	1V	200μA	TO92

(See Figure 4.9 for encapsulations and pin connections)

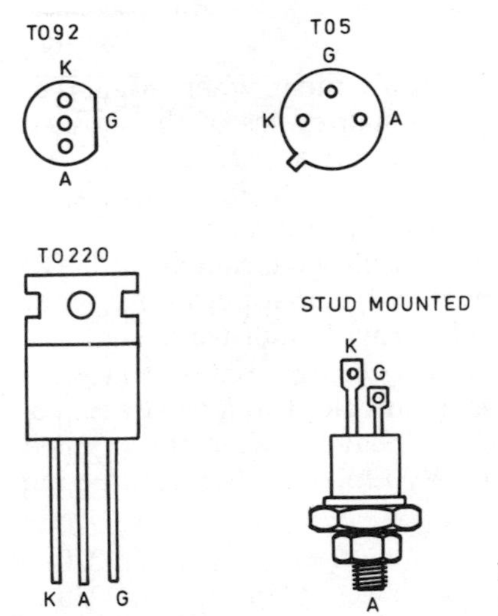

Figure 4.9 Thyristor encapsulations and pin connections

Hints and tips

★ Wherever possible, thyristor trigger pulses should have the fastest possible rise times. Signals with slow rise times or poorly defined edges are generally unsatisfactory for triggering purposes.

★ Sufficient gate current must be made available in order to ensure effective triggering (it is thus necessary to minimise the impedance of the gate driver circuitry as far as possible).

★ Thyristors will turn on faster (and power dissipation within the device will be minimised) as gate current is increased. Care should, however, be taken to ensure that the peak value of gate does not exceed the rated value for the device.
★ The pulse width of the trigger pulse applied to the gate of a thyristor must be kept short in order to minimize gate power dissipation. Negative gate voltages should also be avoided in order to prevent power loss.
★ In order to obtain an adequate range of control in a.c. power control applications, the thyristor triggering circuit should be designed so that it will provide effective triggering over a sufficiently wide angle of the applied a.c. voltage. Failure to observe this rule will result in a limited range of control.

Triacs

Triacs are a refinement of the thyristor which, when triggered, conduct on both positive and negative half-cycles of the applied voltage. Triacs have three terminals known as *main-terminal one (MT1), main terminal two (MT2)* and *gate (G)*, as shown in Figure 4.10. Triacs can be triggered by both positive and negative voltages present at the gate. Triacs thus provide full-wave control and offer superior performance in a.c. power control applications when compared with thyristors which only provide half-wave control.

In order to simplify the design of triggering circuits, triacs are often used in conjunction with *diacs* (equivalent to a bi-directional zener diode). A typical diac conducts heavily when the applied voltage exceeds approximately ±32V. Once in the conducting

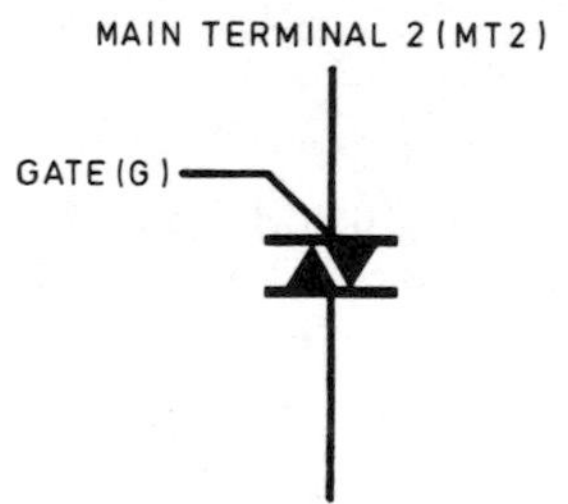

Figure 4.10 Triac connections

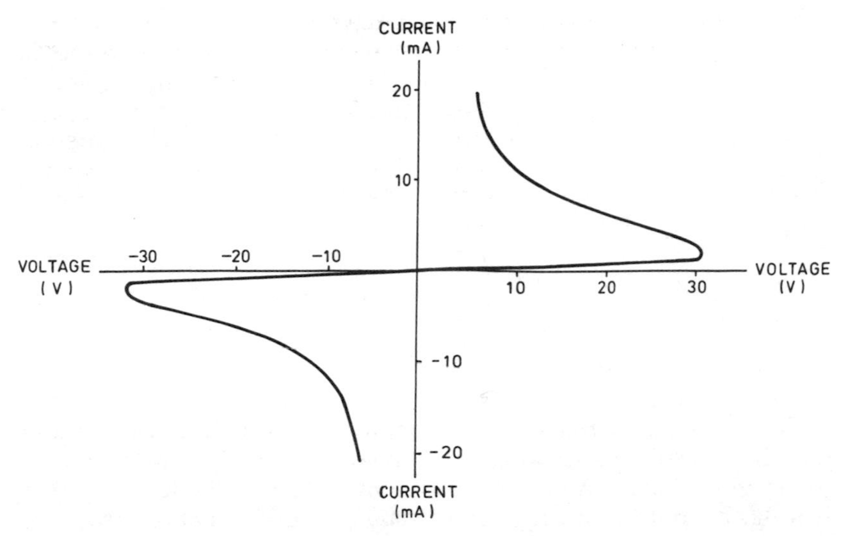

Figure 4.11 Typical diac characteristics

state, the resistance of the diac falls to a very low value and thus a large value of current will flow. The characteristic of a typical diac is shown in Figure 4.11.

Triac data

The following table summarises the characteristics of a variety of popular triacs:

Type	$I_{T(RMS)}$	V_{RRM}	V_{GT}	$I_{GT(TYP)}$	*Case style*
BT139	15A	600V	1.5V	5mA	TO220
TIC206M	4A	600V	2V	5mA	TO220
TIC216M	6A	600V	3V	5mA	TO220
TIC225M	8A	600V	2V	20mA	TO220
TIC226M	8A	600V	2V	50mA	TO220
TIC236M	12A	600V	2V	50mA	TO220
TIC246M	16A	600V	2V	50mA	TO220
TICP206D	1.5A	400V	2.5V	2.5mA	TO92
TICP206M	1.5A	600V	2.5V	2.5mA	TO92

(See Figure 4.12 for encapsulations and pin connections)

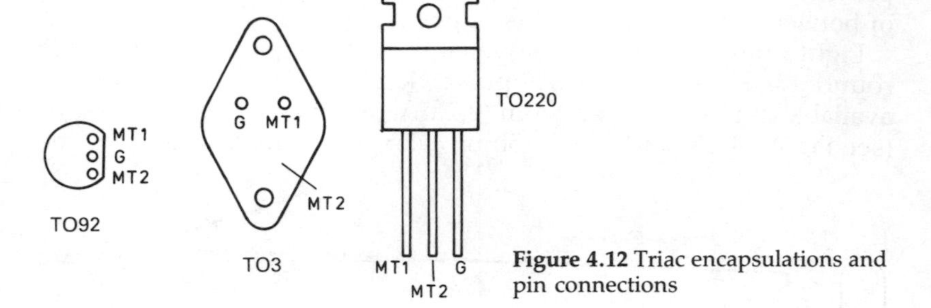

Figure 4.12 Triac encapsulations and pin connections

Hints and tips

★ Thyristors and triacs switch on and off very rapidly. In a.c. power control applications, this rapid switching can result in transients which may be conveyed some distance via the a.c. mains wiring. To minimise such effects and prevent radiation of noise, an L-C filter should be fitted in close proximity to the power control device, as shown in Figure 4.13.

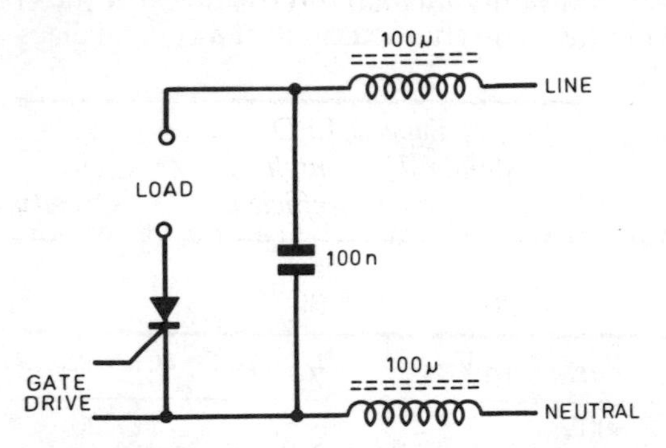

Figure 4.13 Simple power line filter for use with a thyristor or triac power controller

Light emitting diodes

Light emitting diodes (LEDs) can be used as general purpose indicators and, compared with conventional filament lamps, operate from significantly smaller voltages and currents. LEDs are also very much more reliable than filament lamps. Most LEDs will provide a reasonable level of light output when a forward current of between 5mA and 20mA is applied.

Light emitting diodes are available in various formats with the round types being most popular. Round LEDs are commonly available in the 3mm and 5mm (0.2 inch) diameter plastic packages (see Figure 4.14) and also in 5mm × 2mm rectangular format. The

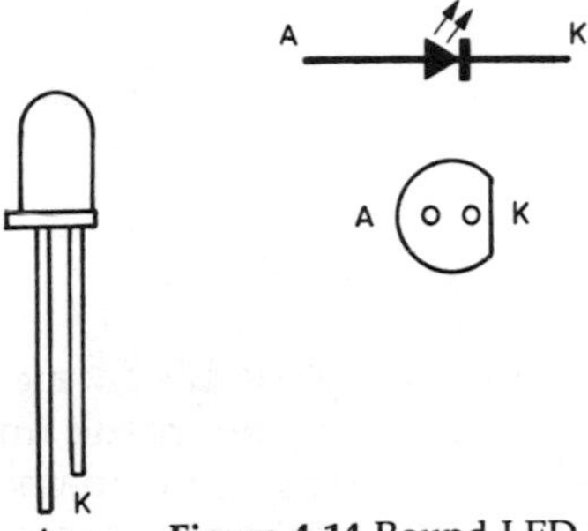

Figure 4.14 Round LED encapsulation

viewing angle for round LEDs tends to be in the region of 20° to 40° whereas, for rectangular types this is increased to around 100°. Typical characteristics for commonly available red LEDs are given below:

Parameter	*Type of LED*			
	standard	*standard*	*high efficiency*	*high intensity*
Diameter (mm)	3	5	5	5
Max. forward current (mA)	40	30	30	30
Typical forward current (mA)	12	10	7	10
Typical forward voltage drop (V)	2.1	2.0	1.8	2.2
Max. reverse voltage (V)	5	3	5	5
Max. power dissipation (mW)	150	100	27	135
Peak wavelength (nm)	690	635	635	635
Typical unit cost	15p	18p	24p	35p

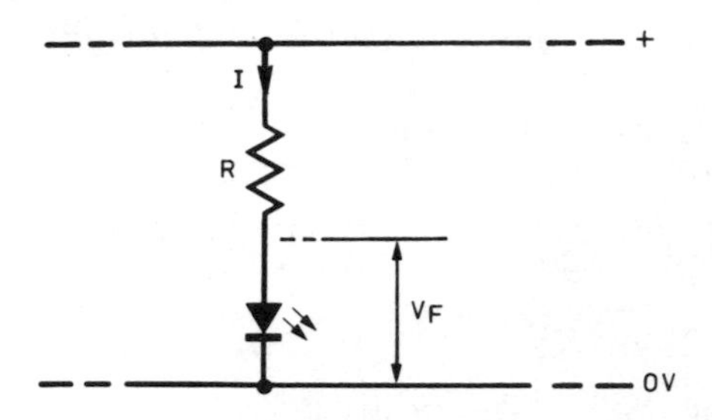

Figure 4.15 Typical LED indicator circuit

In order to limit the forward current to an appropriate value, it is usually necessary to include a fixed resistor in series with a LED indicator, as shown in Figure 4.15. The value of the resistor may be calculated from:

$$R = \frac{V - V_F}{I}$$

where V_F is the forward voltage drop produced by the LED and V is the applied voltage. It is usually safe to assume that V_F will be 2V and choose the nearest preferred value for R. Typical values of LED series resistor are given in the table:

Supply voltage (V)	*Series resistance (Ω)* low power LED (5mA nom)	*standard LED (10mA nom)*	*high power LED (20mA nom)*
3	220	180	56
5	680	270	150
6	820	390	220
9	1.5k	680	390
12	2.2k	1k	560
15	2.7k	1.2k	680
18	3.3k	1.5k	820
24	4.7k	2.2k	1.2k

Example 4.3

An LED is to be used to indicate the presence of a 21V d.c. supply rail. If the LED has a nominal forward voltage of 2.2V, and is rated at a current of 15mA, determine the value of series resistor required.

Here we can use the formula:

$$R = \frac{V - V_F}{I} = \frac{21V - 2.2V}{15mA} = \frac{18.8V}{15mA} = 1.25k\Omega$$

The nearest preferred value is 1.2kΩ. The power dissipated in the resistor will be given by:

$$P = I \times V = 15mA \times 18.8V = 280mW$$

Hence the resistor should be rated at 0.33W, or greater.

Hints and tips

★ Avoid inadvertent reverse LED connection – reverse voltages in excess of about 5V will cause permanent damage!

★ For battery powered equipment (particularly where a number of LED indicators are used) minimal values of forward current should be employed in order to ensure long battery life. A forward current of 5mA (per LED) will be perfectly adequate in many applications.

★ Where several LEDs are to be used together, they should be connected in series (and not in parallel) in order to ensure equal levels of light output.

★ Yellow and green LED generally give less light output (for a given forward current) than their standard red counterparts. To maintain an equal light output when several LEDs of different colours are used together, different values of series resistor may be employed. As a rule of thumb, series resistors for yellow and green LEDs should be chosen so that they are 10% to 15% lower in value than those used with red diodes (care should, however, be taken to ensure that operating currents are still within the manufacturer's specified maximum upper limit).

★ In applications involving low a.c. voltages, a conventional low-current silicon diode (e.g. 1N4148) can be wired in parallel with a LED to provide a simple a.c. indicator (see Figure 4.16).

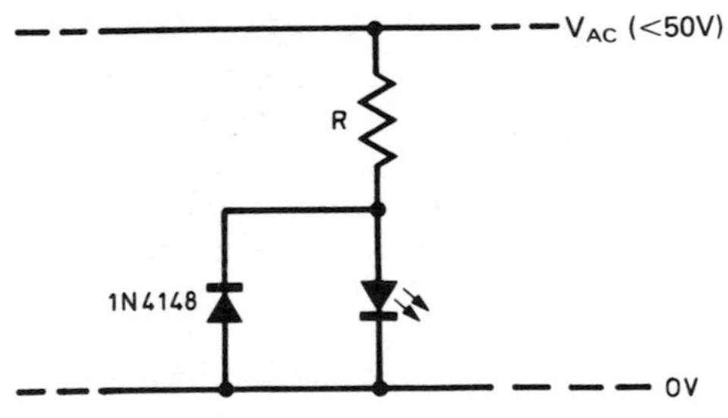

Figure 4.16 LED indicator for a.c. operation

5 Transistors

Transistor types

Transistor is short for *transfer resistor*, a term which provides something of a clue as to how the device operates; the current flowing in one circuit (the *output*) is determined by the current flowing in another (the *input*). Since transistors are three-terminal devices, one electrode must remain common to both the input and the output.

Transistors fall into two main categories (*bipolar* and *field-effect*) and are also classified according to the semiconductor material employed (silicon or germanium) and to their field of application (e.g. general purpose, switching, high-frequency, etc). The following terminology is often used:

Low-frequency
Transistor designed specifically for AF applications (below 100kHz).

High-frequency
Transistors designed specifically for RF applications (100kHz and above).

Power
Transistors which operate at significant power levels (such devices are often sub-divided into audio frequency and radio frequency power types).

Switching
Transistors designed for switching applications.

Low-noise
Transistors which have low-noise characteristics and which are intended primarily for the amplification of low-amplitude signals.

High-voltage
Transistors designed specifically to handle high voltages.

Driver
Transistors which operate at medium power and voltage levels and which are often used to precede a final stage which operates at an appreciable power level.

Transistor coding

The European system for classifying transistors involves an alphanumeric code which employs either two letters and three figures (general purpose transmitters) or three letters and two figures (special purpose transmitters). The first two letters have the following significance:

First letter – semiconductor material:
A germanium
B silicon
Second letter – application:
C low-power, low-frequency
D high-power, low-frequency
F low-power, high-frequency
L high-power, high-frequency

In the case of transistors for specialised applications, the third letter does not generally have any particular significance.

Example 5.6
Identify each of the following transistors:
(i) AF115
(ii) BC108
(iii) BD135
(iv) BFY50

Transistor (i) is a general purpose low-power, high-frequency germanium transistor.

Transistor (ii) is a general purpose low-power, low-frequency silicon transistor.

Transistor (iii) is a general purpose high-power, low-frequency silicon transistor.

Transistor (iv) is a special purpose low-power, high-frequency silicon transistor.

Bipolar transistors

Bipolar transistors generally comprise p-n-p or n-p-n junctions of either silicon (Si) or germanium (Ge) material (see Figure 5.1). In either case the electrodes are labelled collector, base, and emitter (Figure 5.2). Silicon transistors are superior when compared with germanium transistors in the vast majority of applications (particularly at high tempertures) and thus germanium devices are very rarely encountered.

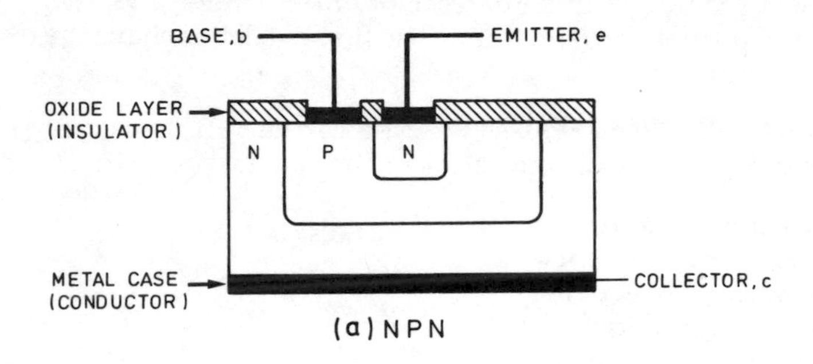

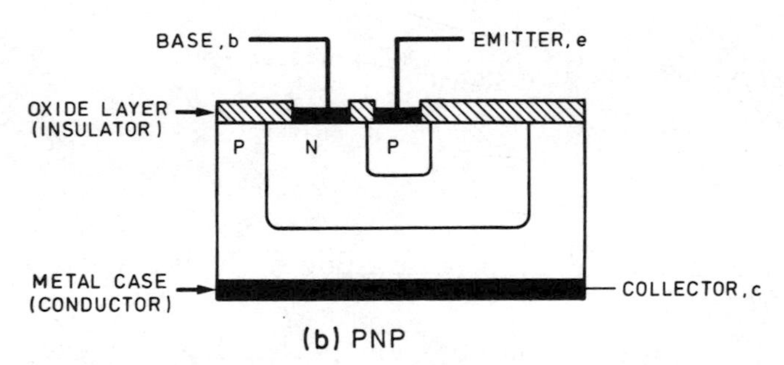

Figure 5.1 NPN and PNP transistors

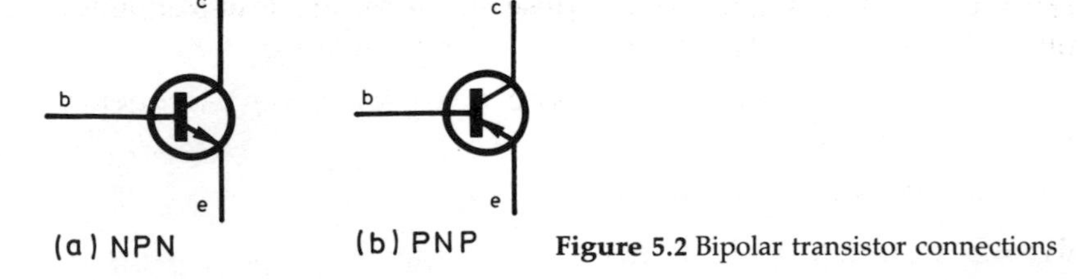

Figure 5.2 Bipolar transistor connections

Each junction within the transistor, whether it be collector-base or base-emitter, constitutes a p-n junction diode. The base region is, however, made very narrow so that carriers are swept across and a relatively small current flows in the base. The current flowing in the emitter circuit is typically 100 times greater than that flowing in the base. The direction of current flow is from emitter to collector in the case of a PNP transistor, and collector to emitter in the case of an NPN device.

The equation which relates current flow in the collector, base, and emitter currents is:

$I_E = I_B + I_C$

where I_E is the emitter current, I_B is the base current, and I_C is the collector current (all expressed in terms of the same units).

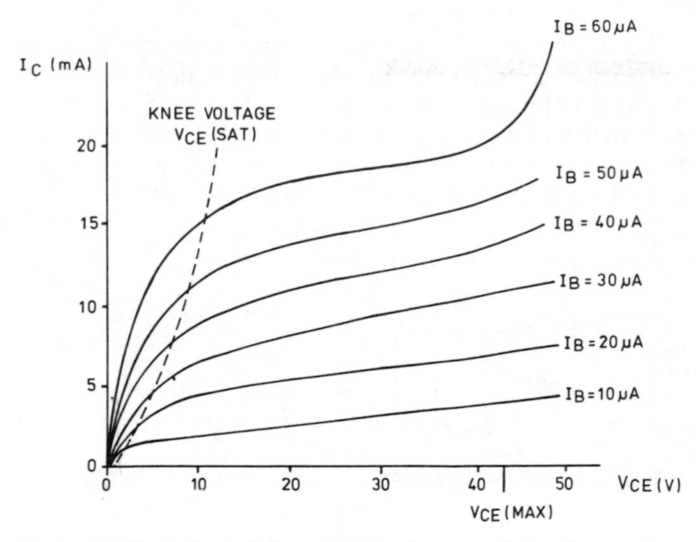

Figure 5.3 Typical set of characteristics for a small-signal general purpose NPN transistor

A typical set of transistor characteristics for a small-signal general purpose NPN transistor are shown in Figure 5.3.

Bipolar transistor parameters

The current gain offered by a transistor is a measure of its effectiveness as an amplifying device. The most commonly quoted parameter is that which relates to common emitter mode. In this mode, the input current is applied to the base and the output current appears in the collector (the emitter is effectively common to both the input and output circuits).

The common emitter current gain is given by:

$$h_{FE} = I_C/I_B$$

where h_{FE} is the hybrid parameter which represents large signal (d.c.) forward current gain, I_C is the collector current, and I_B is the base current. When small (rather than large) signal operation is considered, the values of I_C and I_B are incremental (i.e. small changes rather than static values). The current gain is then given by:

$$h_{fe} = \Delta I_c/\Delta I_b$$

where h_{fe} is the hybrid parameter which represents small signal (a.c.) forward current gain, ΔI_c is the change in collector current which results from a corresponding change in base current, ΔI_b.

Darlington transistors (see Figure 5.4) are a special form of compound bipolar transistor in which a very high value of forward current gain can be achieved (typically several thousands).

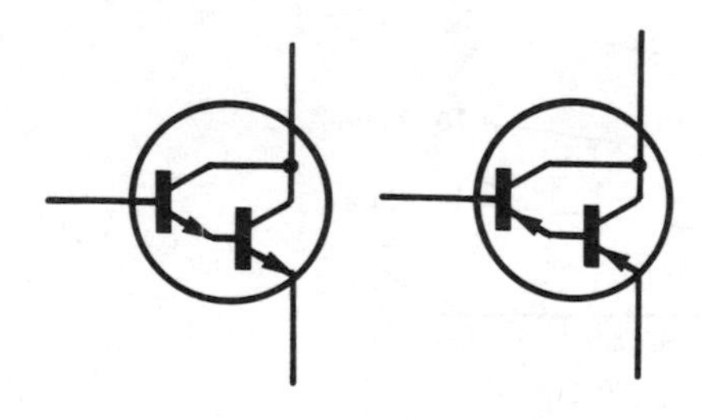

Figure 5.4 Darlington transistors

Other important parameters to note are:

I_c max the maximum value of collector current

V_{ceo} max the maximum value of collector-emitter voltage with the emitter terminal left open-circuit

V_{cbo} max the maximum value of collector-base voltage with the base terminal left open-circuit

P_t max the maximum total power dissipation

h_{fe} the small-signal common-emitter current gain

h_{fe} max the maximum value of small-signal common-emitter current gain

h_{fe} min the minimum value of small-signal common-emitter current gain

f_t typ the transition frequency (i.e. the frequency at which the small signal common-emitter current gain falls to unity)

Bias for bipolar transistors

In most circuits (particualrly those designed for linear operation), a static bias current must be applied to the transistor in order to

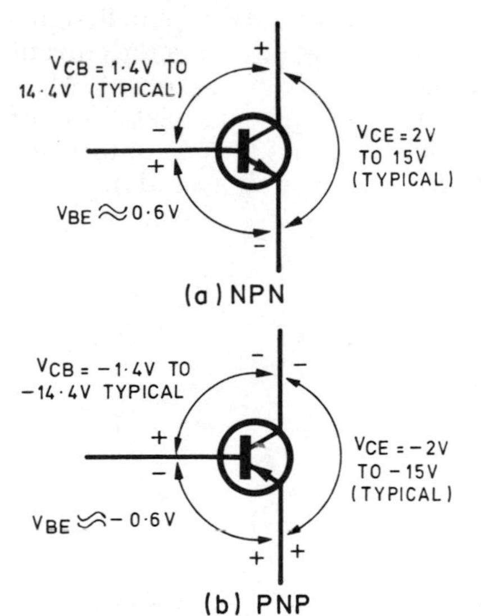

Figure 5.5 Bipolar transistor potentials when biased for linear operation

obtain satisfactory operation. This bias is usually applied in the form of a small current to the base terminal. This current sets up a standing (quiescent) current of larger magnitude in the collector circuit.

The voltage developed across a forward biased base-emitter junction of a silicon transistor will be 0.6V (0.1V for a germanium device). In the case of an NPN device, the base and collector will be positive with respect to the emitter whilst for a PNP device the base and collector will be negative with respect to the emitter (see Figure 5.5).

Bipolar transistor data

Bipolar transistor summary

Device	*Type*	*Application*	*Equivalent*	*Near equivalent*	*Complement*	*Case style*
2N2219A	NPN	Switching		2N2222A (TO18A case)		TO39
2N2222A	NPN	Switching		2N2219A (TO39 case)		TO18A
2N2369A	NPN	RF				TO18A
2N2905	PNP	Switching				TO5
2N2926	NPN	General purpose				TO98
2N3053	NPN	Driver				TO5
2N3054	NPN	Medium power				TO66
2N3055	NPN	High power		TIP3055 (plastic case)	MJ2955	TO3A
2N3702	PNP	General purpose		BC558 (TO92B case)		TO92A
2N3703	PNP	General purpose		2N3702		TO92A
2N3866	NPN	RF				TO5
2N3903	NPN	Switching		2N3904		TO92C
2N3904	NPN	Switching		2N3903	2N3906	TO92C
2N3906	PNP	Switching			2N3904	TO92C
2N4427	NPN	RF				TO5
2N5401	PNP	High voltage				TO92C
2N706	NPN	Switching		2N2369A, BSX20		TO18A
BC107	NPN	Driver	BC237 (plastic)		BC177	TO18A
BC108	NPN	General purpose	BC238 (plastic)	BC109	BC178	TO18A
BC109	NPN	Small-signal	BC239 (plastic)	BC108	BC179	TO18A
BC117	NPN	High voltage				TO39
BC142	NPN	Driver			BC143	TO39

Device	*Type*	*Application*	*Equivalent*	*Near equivalent*	*Complement*	*Case style*
BC143	PNP	Driver			BC142	TO39
BC177	PNP	Driver			BC107	TO18A
BC178	PNP	General purpose		BC179	BC108	TO18A
BC179	PNP	Small-signal		BC179	BC109	TO18A
BC182L	NPN	General purpose		BC184L	BC212L	TO92A
BC184L	NPN	General purpose		BC183L	BC214L	TO92A
BC212L	PNP	General purpose		BC214L	BC182L	TO92A
BC214L	PNP	General purpose		BC213L	BC184L	TO92A
BC237	PNP	Driver			BC337	TO92B
BC337	NPN	Driver			BC327	TO92B
BC441	NPN	Driver			BC461	TO39
BC461	PNP	Driver			BC441	TO39
BC478	PNP	General purpose		BC178, BC479		TO18A
BC547	NPN	Driver		BC107 (metal case)	BC557	TO92B
BC548	NPN	General purpose		BC549, BC108 (metal case)	BC558	TO92B
BC549	NPN	Small-signal		BC548, BC109 (metal case)	BC559	TO92B
BC557	PNP	Driver		BC177 (metal case)	BC547	TO92B
BC558	PNP	General purpose	BC559	BC178 (metal case)	BC548	TO92B
BC559	PNP	Small-signal		BC558, BC179 (metal case)	BC549	TO92B
BCY70	PNP	General purpose		BCY71		TO18A
BD131	NPN	Medium power			BD132	TO126m
BD132	PNP	Medium power			BD131	TO126m
BF180	NPN	RF				TO72A
BF258	NPN	High voltage		BF259		TO5
BF259	NPN	High voltage		BF258		TO5

Device	*Type*	*Application*	*Equivalent*	*Near equivalent*	*Complement*	*Case style*
BF337	NPN	High voltage				TO39
BFY50	NPN	Driver		BFY51		TO39
BFY51	NPN	General purpose		BFY50, BFY52		TO39
BFY52	NPN	General purpose		BFY51		TO39
BFY90	NPN	RF				TO72B
BSX20	NPN	Switching	2N2369A			TO18A
MJ11015	PNP	Power Darlington			MJ11016	TO3A
MJ11016	NPN	Power Darlington			MJ11015	TO3A
MJ2501	PNP	Power Darlington			MJ3001	TO3A
MJ2955	PNP	High power		TIP2955 (plastic case)	2N3055	TO3A
MJ3001	NPN	Power Darlington			MJ2501	TO3A
TIP121	NPN	Power Darlington			TIP126	TO220A
TIP126	PNP	Power Darlington			TIP121	TO220A
TIP132	NPN	Power Darlington			TIP137	TO220A
TIP137	PNP	Power Darlington			TIP132	TO220A
TIP141	NPN	Power Darlington			TIP146	TAB-A
TIP146	PNP	Power Darlington			TIP141	TAB-A
TIP2955	PNP	High power		MJ2955 (metal case)	TIP3055	TAB-A
TIP3055	NPN	High power		2N3055 (metal case)	TIP2955	TAB-A
TIP31A	NPN	Power		TIP31C	TIP32A	TO220A
TIP32A	PNP	Power		TIP32C	TIP31A	TO220A
TIP41A	NPN	Power		TIP41C	TIP42A	TO220A
TIP42A	PNP	Power		TIP42C	TIP41A	TO220A
ZTX108	NPN	General purpose		BC108 (metal case)		E-line
ZTX300	NPN	General purpose		ZTX302	ZTX500	E-line
ZTX500	PNP	General purpose		ZTX502	ZTX300	E-line

General purpose NPN transistors

Device	*Type*	I_c *max.*	V_{ceo} *max.*	V_{cbo} *max.*	P_t *max.*	h_{fe}	*at* I_c	f_t *typ.*
2N2926	NPN	100mA	18V	18V	200mW	200	2mA	200MHz
BC10B	NPN	100mA	20V	30V	300mW	125	2mA	250MHz
BC182L	NPN	200mA	50V	60V	300mW	225	2mA	150MHz
BC184L	NPN	200mA	30V	45V	300mW	250	2mA	150MHz
BC548	NPN	100mA	30V	30V	625mW	250	2mA	300MHz
BFY51	NPN	1A	30V	60V	800mW			50MHz
BFY52	NPN	1A	20V	40V	800mW			50MHz
ZTX10B	NPN	100mA	20V	30V	200mW	300	2mA	300MHz
ZTX300	NPN	500mA	25V	25V	300mW	150	10mA	150MHz

General purpose PNP transistors

Device	*Type*	I_c *max.*	V_{ceo} *max.*	V_{cbo} *max.*	P_t *max.*	h_{fe}	*at* I_c	f_t *typ.*
2N3702	PNP	200mA	25V	40V	360mW			100MHz
2N3703	PNP	500mA	30mV	50V	300mW	125	2mA	100MHz
BC178	PNP	100mA	25V	30V	300mW	125	2mA	200MHz
BC212L	PNP	200mA	50V	60V	300mW	200	2mA	200MHz
BC214L	PNP	200mA	30V	45V	300mW	225	2mA	200MHz
BC478	PNP	50mA	40V	40V	360mW	175	2mA	150MHz
BC558	PNP	100mA	45V	50V	625mW	250	2mA	360MHz
BCY70	PNP	200mA	40V	50V	360mW	150	2mA	200MHz
ZTX500	PNP	500mA	25V	25V	300mW	150	10mA	150MHz

Small-signal transistors

Device	*Type*	I_c *max.*	V_{ceo} *max.*	P_t *max.*	h_{fe}	*at* I_c	f_t *typ.*
BC109	NPN	100mA	20V	360mW	250	2mA	250MHz
BC179	PNP	100mA	20V	300mW	240	2mA	200MHz
BC549	NPN	100mA	30V	625mW	250	2mA	300MHz
BC559	PNP	100mA	30V	625mW	250	2mA	250MHz

Switching transistors

Device	*Type*	I_c *max.*	V_{ceo} *max.*	V_{cbo} *max.*	P_t *max.*	h_{fe} *min.*	h_{fe} *max.*	f_t *typ.*
2N2219A	NPN	800mA	40V	75V	800mW	75		300MHz
2N2222A	NPN	800mA	40V	75V	500mW	35		250MHz
2N2905	PNP	600mA	40V	60V	600mW	150	300	200MHz
2N3903	NPN	200mA	40V	60V	350mW	50	150	250MHz
2N3904	NPN	200mA	40V	60V	310mW	100	300	300MHz
2N3906	PNP	200mA	40V	40V	310mW	100	300	250MHz
2N706	NPN	20mA	20V	25V	300mW			200MHz
BSX20	NPN	500mA	15V	40V	360mW			500MHz

RF transistors

Device	*Type*	I_c *max.*	V_{ceo} *max.*	P_t *max.*	h_{fe}	*at* I_c	f_t *typ.*
2N2369A	NPN	200mA	15V	360mW			500MHz
2N3866	NPN	400mA	30V	3W	105	50mA	700MHz
2N4427	NPN	500mA	20V	2.5W	100	100mA	500MHz
BF180	NPN	20mA	20V	150mW	100	10mA	650MHz
BFY90	NPN	50mA	15V	200mW			1.4GHz

Driver transistors

Device	*Type*	I_c *max.*	V_{ceo} *max.*	V_{cbo} *max.*	P_t *max.*	h_{fe} *min.*	h_{fe} *max.*	f_t *typ.*
2N3053	NPN	700mA	40V	60V	800mW	125	250	100MHz
BC107	NPN	100mA	45V	50V	360mW	110	450	250MHz
BC142	NPN	800mA	60V	80V	800mW	20	250	80MHz
BC143	PNP	800mA	60V	60V	800mW	25	250	160MHz
BC177	PNP	100mA	45V	50V	300mW	125	500	200MHz
BC327	PNP	500mA	45V	50V	625mW	100	600	260MHz
BC337	NPN	500mA	45V	50V	625mW	100	600	200MHz
BC441	NPN	1A	60V	75V	1W	40	250	50MHz
BC461	PNP	1A	60V	75V	1W	40	250	50MHz
BC547	NPN	100mA	45V	50V	625mW	110	800	300MHz
BC557	PNP	100mA	45V	50V	625mW	110	800	320MHz
BFY50	NPN	1A	35V	80V	800mW	30		60MHz

Power transistors

Device	*Type*	*I_c max.*	*V_{ceo} max.*	*V_{cbo} max.*	*P_t max.*	*h_{fe} min.*	*h_{fe} max.*	*f_t typ.*
2N3054	NPN	4A	55V	90V	29W	25		1MHz
2N3055	NPN	15A	60V	100V	115W	20	70	1MHz
BD131	NPN	3A	45V	70V	15W	20	150	60MHz
BD132	PNP	3A	45V	45V	15W	20	60	60MHz
MJ2955	PNP	15A	60V	100V	150W	20	70	4MHz
TIP2955	PNP	15A	60V	100V	90W	5	30	8MHz
TIP3055	NPN	15A	60V	100V	90W	5	30	8MHz
TIP31A	NPN	3A	60V	60V	40W	10	60	8MHz
TIP32A	PNP	3A	60V	60V	40W	10	40	8MHz
TIP41A	NPN	6A	60V	60V	65W	15		3MHz
TIP42A	PNP	6A	60V	60V	65W	15		3MHz

Darlington transistors

Device	*Type*	*I_c max.*	*V_{ceo} max.*	*V_{cbo} max.*	*P_t max.*	*h_{fe} min.*	*f_t typ.*
MJ11015	PNP	30A	120V	120V	200W	2000	
MJ11016	NPN	30A	120V	120V	200W	2000	
MJ2501	PNP	10A	80V	80V	150W	1000	1MHz
MJ3001	NPN	10A	80V	80V	150W	1000	1MHz
TIP121	NPN	5A	80V	80V	65W	1000	1MHz
TIP126	PNP	5A	80V	80V	65W	1000	1MHz
TIP132	NPN	8A	100V	100V	70W	1000	1MHz
TIP137	PNP	8A	100V	100V	70W	1000	1MHz
TIP141	NPN	10A	80V	80V	125W	1000	1MHz
TIP146	PNP	10A	80V	80V	125W	1000	1MHz

Example 5.1
A transistor operates with a collector current of 97mA and an emitter current of 98mA. Determine the value of base current and common-emitter current gain.

Since $I_E = I_B + I_C$, the base current will be given by:

$I_B = I_E - I_C = 98mA - 97mA = 1mA$

The common-emitter current gain will be given by:

$h_{FE} = I_C/I_B = 97mA/1mA = 97$

Example 5.2
A transistor is to be used in a regulator circuit in which a collector current of 1.5A is to be controlled by a base current of 30mA. What value of h_{FE} will be required?

The required current gain can be found from:

$h_{FE} = I_C/I_B = 1.5A/30mA = 1500mA/30mA = 50$

Example 5.3
A transistor is used in a linear amplifier arrangement. The transistor has small and large signal current gains of 175 and 200 respectively and bias is arranged so that the static value of collector current is 10mA. Determine the value of base bias current and the change of output (collector) current that would result from a 10μA change in input (base) current.

The value of base bias current can be determined from:

$I_B = I_C/h_{FE} = 10mA/200 = 50\mu A$.

The change of collector current resulting from a 10μA change in input current will be given by:

$I_c = h_{fe} \times I_b = 175 \times 10\mu A = 1.75mA$

Hints and tips

★ Current gain (h_{fe}) varies with collector current. For most small-signal transistors, h_{fe} is a maximum at a collector current in the range 1mA and 10mA. h_{fe} falls to very low values for power transistors (other than Darlington devices) when operating at very high values of collector current.

★ Most transistor parameters (particularly common-emitter current gain, h_{fe}) are liable to wide variation from one device to the next. It is, therefore, important to design circuits on the basis of the *minimum* value for h_{fe} in order to ensure successful operation with a variety of different devices.

★ Transistors will usually operate reliably as amplifiers using conventional circuits in common-emitter mode at frequencies which are not in excess of one tenth of the quoted value for f_t. Special techniques (including neutralisation or the use of common-base mode) are required when it is necessary to operate at frequencies in excess of this value.

★ Power transistors should be de-rated at high operating temperatures (particularly when heat sinking arrangements do not meet the manufacturer's recommendations). The normal requirement is to de-rate power dissipation linearly to zero at 100°C whenever the junction temepratture exceeds 40°C.

Field effect transistors

Field effect transistors (FETs) comprise a channel of p or n-type material surrounded by material of the opposite polarity. The ends of the channel (in which conduction takes place) form electrodes known as the *source* and *drain*. The effective width of the channel (in which conduction takes place) is controlled by a charge placed on the third (*gate*) electrode. The effective resistance between the source and drain is thus determined by the voltage present at the gate.

Field effect transistors are available in two basic forms; *junction gate* and *insulated gate* (Figure 5.6). The gate-source junction of a junction gate field effect transistor (JFET) is effectively a reverse-biased p-n junction. The gate connection of an insulated gate field effect transistor (IGFET), on the other hand, is insulated from the channel and charge is capacitively coupled to the channel. IGFETs use either metal on silicon (MOS) or silicon on sapphire (SOS) technology.

JFET devices are less noisy and more stable than comparable IGFET devices. JFET devices offer source input impedances of around 100MΩ compared with 10000MΩ for comparable IGFETs. (Note, however, that FET devices offer very much higher input impedances (at the source) than bipolar transistors (at the base).) IGFET devices generally offer improved switching characteristics as they combine low drain-source resistance in the on-state with very high drain-source resistance in the off-state.

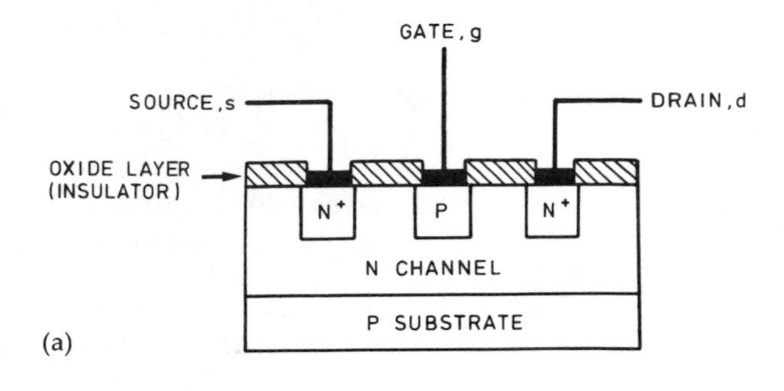

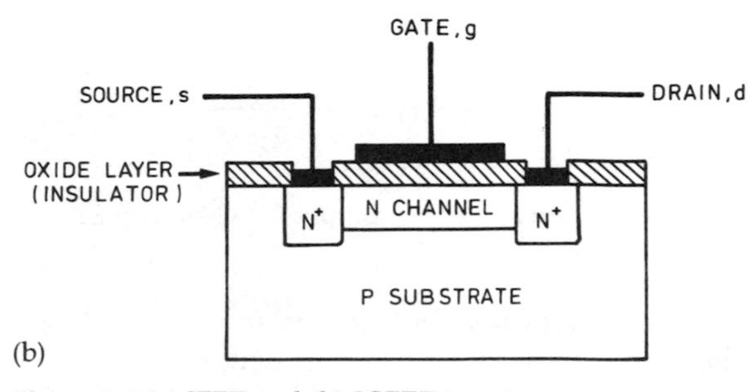

Figure 5.6 (a) JFET and (b) IGFET transistors

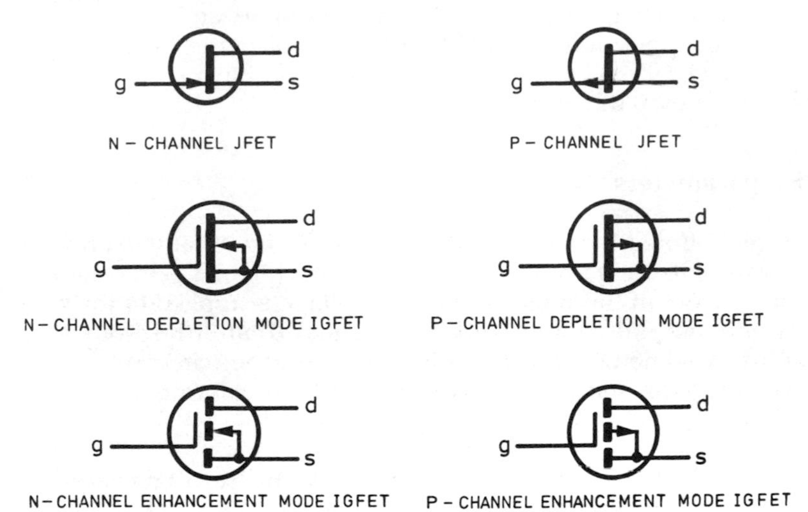

Figure 5.7 Symbols and connections for various types of FET

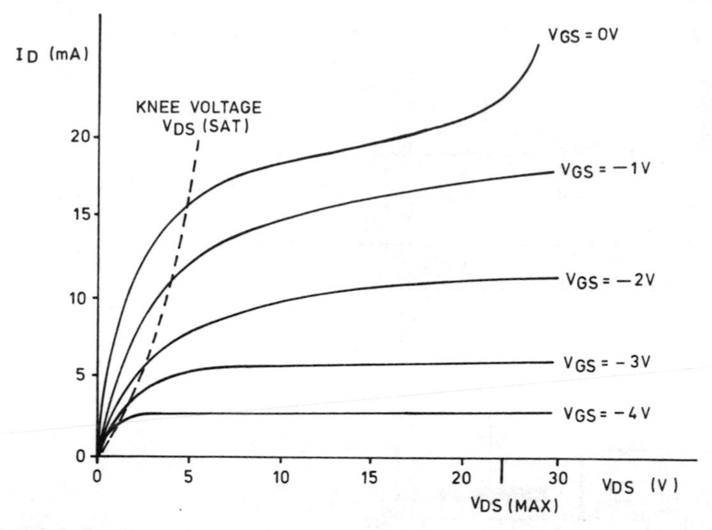

Figure 5.8 Typical set of characteristics for a general purpose N-channel JFET.

IGFETs may be designed for either *depletion mode* or *enhancement mode* operation. In the former case, conduction occurs within the channel even when the gate-source voltage (V_{GS}) is zero. In the latter case, a gate-source bias voltage must be applied in order to obtain conduction within the channel. Symbols and connections for various types of JFET and IGFET are depicted in Figure 5.7.

A typical set of characteristics for a general purpose N-channel JFET are shown in Figure 5.8.

FET parameters

The *gain* offered by a field effect transistor is normally expressed in terms of its forward transfer conductance (g_{fs} or Y_{fs}) in common source mode. In this mode, the input voltage is applied to the gate and the output current appears in the drain (the source is effectively common to both the input and output circuits).

The common source forward transfer conductance is given by:

$$g_{fs} = \Delta I_d / \Delta V_{gs}$$

where ΔI_d is the change in drain current resulting from a corresponding change in gate-source voltage (ΔV_{gs}). The units of forward transfer conductance are siemen (S).

Other important parameters to note are:

I_D max	the maximum drain current
V_{DS} max	the maximum drain-source voltage
V_{GS} max	the maximum gate-source voltage
P_D max	the maximum drain power dissipation
t_r typ	the typical output rise-time in response to a perfect rectangular pulse input
t_f typ	the typical output fall-time in response to a perfect rectangular pulse input
$R_{DS(on)}$ max	the maximum value of resistance between drain and source when the transistor is in the conducting (on) state

Bias for field effect transistors

As with bipolar devices, linear FET applications necessitate the application of bias. The method of applying bias will differ according to the mode of operation (depletion or enhancement) but, in either case, it will involve the application of a gate-source bias voltage. A standing (*quiescent*) value of drain current will result. Typical values of gate-source bias voltage vary from between −2.5V and +2.5V, according to mode of operation.

Field effect transistor data

Field effect transistor summary

Device	*Type*	*Application*	*Near equiv.*	*Complement*	*Case style*
2SJ50	P-chan. MOSFET	Audio power	2SJ56	2SK135	TO3B
2SJ56	P-chan. MOSFET	Audio power	2SJ50	2SK176	TO3B
2SK135	N-chan. MOSFET	Audio power	2SK176	2SK50	TO3B
2SK176	N-chan. MOSFET	Audio power	2SK135	2SK56	TO3B
2N3819	N-chan. JFET	General purpose		2N3820	TO92D
2N3820	P-chan. JFET	General purpose		2N3819	TO92D
2N4092	N-chan. JFET	Switching			TO18B
2N4118	N-chan. JFET	General purpose			TO72D
2N4220	N-chan. JFET	Low-noise			TO72C
2N4351	N-chan. JFET	Switching			TO72D
2N4391	N-chan. JFET	Switching	2N4392		TO18B
2N4392	N-chan. JFET	Switching	2N4391, 2N4393		TO18B
2N4393	N-chan. JFET	Switching	2N4392		TO18B
2N4416	N-chan. JFET	RF amplifier			TO72C
2N4858	N-chan. JFET	Switching			TO18D
2N4861	N-chan. JFET	Switching			TO18C
2N5457	N-chan. JFET	General purpose			TO92E

continued overleaf

Device	*Type*	*Application*	*Near equiv.*	*Complement*	*Case style*
2N5460	P-chan. JFET	Low-noise	2N5461		TO92F
2N5461	P-chan. JFET	Low-noise	2N5460		TO92E
2N5486	N-chan. JFET	RF amplifier			TO92E
2N7000	N-chan. MOSFET	Switching			TO92D
2N7007	N-chan. MOSFET	Switching			TO92D
2N7010	N-chan. MOSFET	General purpose			TO237D
2N7014	N-chan. MOSFET	General purpose			TO220AB
2N7054	N-chan. MOSFET	General purpose			TO218AB
2N7055	N-chan. MOSFET	General purpose			TO218AB
2N7058	N-chan. MOSFET	General purpose			TO218AB
BF244A	N-chan. JFET	RF amplifier	BF245A		TO92D
BF245A	N-chan. JFET	RF amplifier	BF244A		TO92E
BUZ11	N-chan. MOSFET	Switching			TO220AB
IRF120	N-chan. MOSFET	General purpose			TO3C
IRF130	N-chan. MOSFET	General purpose			TO3C
IRF330	N-chan. MOSFET	General purpose			TO3C
IRF510	N-chan. MOSFET	General purpose	IRF511		TO220AB
IRF511	N-chan. MOSFET	General purpose	IRF510		TO220AB
IRF520	N-chan. MOSFET	General purpose			TO220AB
IRF530	N-chan. MOSFET	General purpose	IRF531		TO220AB
IRF531	N-chan. MOSFET	General purpose	IRF530		TO220AB
IRF540	N-chan. MOSFET	General purpose			TO220AB
IRF610	N-chan. MOSFET	General purpose			TO220AB
IRF620	N-chan. MOSFET	General purpose	IRF621		TO220AB
IRF621	N-chan. MOSFET	General purpose	IRF620		TO220AB
IRF640	N-chan. MOSFET	General purpose			TO220AB
IRF710	N-chan. MOSFET	General purpose			TO220AB
IRF720	N-chan. MOSFET	General purpose			TO220AB
IRF730	N-chan. MOSFET	General purpose	IRF720		TO220AB
IRF830	N-chan. MOSFET	General purpose	IRF831		TO220AB
IRF831	N-chan. MOSFET	General purpose	IRF830		TO220AB
IRF840	N-chan. MOSFET	General purpose			TO220AB
J309	N-chan. JFET	RF amplifier	J310		TO92E
J310	N-chan. JFET	RF amplifier	J309		TO92E
VN10LM	N-chan. MOSFET	General purpose	VN10KM		TO237D
VN10KM	N-chan. MOSFET	General purpose	VN10LM		TO237D
VN46AF	N-chan. MOSFET	General purpose			TO202
VN66AF	N-chan. MOSFET	General purpose			TO202
VN88AF	N-chan. MOSFET	General purpose			TO202

Audio power MOSFETs

Device	*Type*	I_D *max*	V_{DS} *max*	P_D *max*	g_{fs} *min*
2SJ50	P-chan.	7A	160V	100W	0.7S
2SJ56	P-chan.	8A	200V	120W	0.7S
2SK135	N-chan.	7A	160V	100W	0.7S
2SK176	N-chan.	8A	200V	120W	0.7S

General purpose MOSFETs

Device	*Type*	I_D *max*	V_{DS} *max*	P_D *max*	g_{fs} *min*	$R_{DS(ON)}$ *max*
2N7000	N-chan.	280mA	60V	800mW	0.1S	5Ω
2N7007	N-chan.	87mA	240V	800mW	30mS	45Ω
2N7010	N-chan.	1.3A	60V	1.2W		0.35Ω
2N7014	N-chan.	3.5A	100V	20W	0.75S	0.8Ω
2N7054	N-chan.	38A	100V	150W	8S	0.06Ω
2N7055	N-chan.	28A	200V	150W	8S	0.1Ω
2N7058	N-chan.	12A	500V	150W	6S	0.45Ω
BUZ11	N-chan.	30A	50V	75W	4S	0.04Ω
IRF120	N-chan.	8A	100V	40W	1.5S	0.3Ω
IRF130	N-chan.	14A	100V	75W	4S	0.18Ω
IRF330	N-chan.	5.5A	400V	75W	3S	1Ω
IRF510	N-chan.	4A	100V	20W	1S	0.6Ω
IRF511	N-chan.	4A	60V	20W	1S	0.6Ω
IRF520	N-chan.	8A	100V	40W	1.5S	0.3Ω
IRF530	N-chan.	14A	100V	75W	4S	0.18Ω
IRF531	N-chan.	14A	60V	75W	4S	0.18Ω
IRF540	N-chan.	27A	100V	125W	6S	0.09Ω
IRF610	N-chan.	2.5A	200V	20W	0.8S	1.5Ω
IRF620	N-chan.	5A	200V	40W	1.3S	0.8Ω
IRF621	N-chan.	5A	150V	40W	1.3S	0.8Ω
IRF640	N-chan.	18A	200V	125W	6S	0.18Ω
IRF710	N-chan.	1.5A	400V	20W	0.5S	3.6Ω
IRF720	N-chan.	3A	400V	40W	1S	1.8Ω
IRF730	N-chan.	5.5A	400V	75W	3S	1.5Ω
IRF830	N-chan.	4.5A	500V	75W	2.5S	1.5Ω
IRF831	N-chan.	4.5A	450V	75W	2.5S	1.5Ω
IRF840	N-chan.	8A	500V	125W	4S	0.85Ω
VN10LM	N-chan.	0.3A	60V	1W	0.2S	5Ω
VN10KM	N-chan.	0.3A	60V	1W	0.1S	
VN46AF	N-chan.	2A	40V	15W	0.15S	3Ω
VN66AF	N-chan.	1.8A	60V	15W	0.2S	3Ω
VN88AF	N-chan.	1.6A	80V	15W	0.2S	4Ω

General purpose JFETs

Device	*Type*	I_D *max*	V_{DS} *max*	P_D *max*	g_{fs} *min*
2N3819	N-chan.		25V	200mW	2mS
2N3820	P-chan.		25V	200mW	0.8mS
2N4118	N-chan.		40V	300mW	
2N5457	N-chan.	10mA	25V	310mW	1mS

Switching JFETs

Device	*Type*	V_{DS} *max*	V_{GS} *max*	P_D *max*	t_r *typ*	t_f *typ*
2N4092	N-chan.	40V	−7V	1.8W	20ns	
2N4351	N-chan.	30V	−5V	300mW	65ns	100ns
2N4391	N-chan.	40V	−10V	1.8W	5ns	15ns
2N4392	N-chan.	40V	−5V	1.8W	5ns	20ns
2N4393	N-chan.	40V	−3V	1.8W	5ns	20ns
2N4858	N-chan.	40V	−4V	1.8W	5ns	
2N4861	N-chan.	30V	−4V	360mW	10ns	

JFET RF and low-noise amplifiers

Device	*Type*	I_D *max*	V_{DS} *max*	P_D *max*	g_{fs} *min*
2N4220	N-chan.	15mA	30V	300mW	1mS
2N4416	N-chan.		35V	300mW	4mS
2N5486	N-chan.		25V	310mW	4mS
BF244A	N-chan.	100mA	30V	360mW	3mS
BF245A	N-chan.	100mA	30V	360mW	3mS
J309	N-chan.		25V	360mW	10mS
J310	N-chan.		25V	360mW	8mS

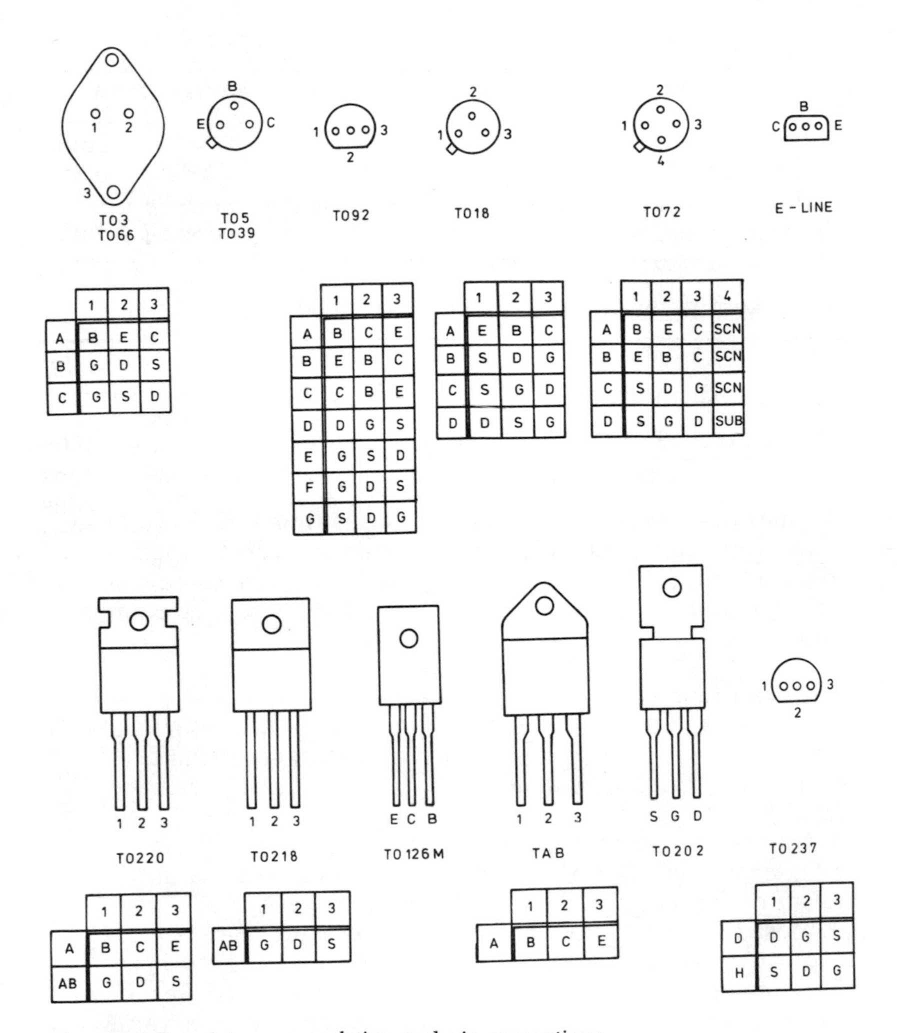

	1	2	3
A	B	E	C
B	G	D	S
C	G	S	D

	1	2	3
A	B	C	E
B	E	B	C
C	C	B	E
D	D	G	S
E	G	S	D
F	G	D	S
G	S	D	G

	1	2	3
A	E	B	C
B	S	D	G
C	S	G	D
D	D	S	G

	1	2	3	4
A	B	E	C	SCN
B	E	B	C	SCN
C	S	D	G	SCN
D	S	G	D	SUB

	1	2	3
A	B	C	E
AB	G	D	S

	1	2	3
AB	G	D	S

	1	2	3
A	B	C	E

	1	2	3
D	D	G	S
H	S	D	G

Figure 5.9 Transistor encapsulation and pin connections

Example 5.4

A FET operates with a drain current of 100mA and a gate-source bias of $-1V$. If the device has a g_{fs} of $0.25S$, determine the change in drain current if the bias voltage increases to $-1.1V$.

The change in gate-source voltage (V_{gs}) is $-0.1V$ and the resulting change in drain current can be determined from:

$I_d = g_{fs} \times V_{gs} = 0.25S \times -0.1V = -0.025A = -25mA$

The new value of drain current will thus be (100mA−25mA) or 75mA.

Example 5.5
A MOSFET is to be used in a static switching application in which a maximum current of 12A is to be controlled. If the device is rated at 75W, determine the maximum permissible value of r_{ds} (on).

The maximum dissipation can be determined from:

$P_D \max = I_L^2 \times r_{ds}$

Thus the maximum value of r_{ds} will be given by:

$$r_{ds} = \frac{P_D \max}{I_L^2} = \frac{75W}{(12)^2} = \frac{75}{144} = 0.52\Omega$$

The maximum permissible value for r_{ds} is thus 0.52Ω. To ensure reliable operation and to avoid the risk of over-dissipation, we should obtain a device which has an 'on' resistance somewhat less than this value (e.g. 0.2Ω, or less). An IRF530 would be eminently suitable.

Hints and tips
★ Care should be taken when handling and soldering MOSFET devices as they can be easily damaged by stray static charges. JFET devices are more robust and are thus less easily damaged by careless handling.
★ Forward transfer conductance (g_{fs}) varies with drain current collector current. For most small-signal devices, g_{fs} is quoted for values of drain current between 1mA and 10mA.
★ Most FET parameters (particularly forward transfer conductance) are liable to wide variation from one device to the next. It is, therefore, important to design circuits on the basis of the *minimum* value for g_{fs} to ensure successful operation with a variety of different devices.
★ Power MOSFETs should be de-rated at high operating temperatures (particularly when heat sinking arrangements do not meet the manufacturer's recommendations). The normal requirement is to de-rate power dissipation linearly to zero at 100°C whenever the junction temperature exceeds 40°C.

6 Integrated circuits

Integrated circuit types

Integrated circuits are complex circuits fabricated on a small slice of silicon. Integrated circuits may contain as few as 10 or as many as 100000 active devices (transistors and diodes) and, with the exception of a few specialised applications (such as amplification at high power levels), integrated circuits have largely rendered discrete circuitry obsolete.

Integrated circuits may be divided into two general classes *linear* (analogue) and *digital*. Typical examples of linear integrated circuits are *operational amplifiers* whereas typical examples of digital integrated circuits are *logic gates*. There are also a number of devices which bridge the gap between the analogue and digital world. Analogue to digital converters (ADC) and digital to analogue converters (DAC) are obvious examples, as is the ubiquitous 555 timer. This latter device contains two operational amplifier stages (configured as voltage comparators) together with a digital bistable stage, a buffer amplifier and an open-collector transistor.

Encapsulation

The most popular forms of encapsulation used for integrated circuits is the dual-in-line (DIL) package which may be fabricated from either plastic or ceramic material (with the latter using a glass hemetic sealant). Common DIL packages have 8, 14, 16, 28, and 40 pins on a 0.1 inch matrix, the pin numbering conventions for which are shown in Figure 6.1.

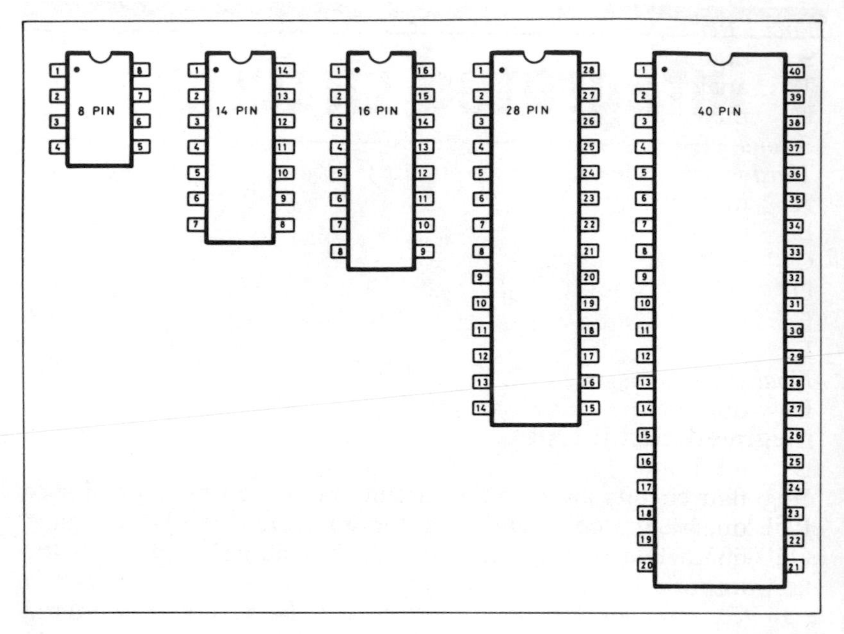

Figure 6.1 Common DIL packages showing pin numbering. (Note that these are shown viewed from the *top*)

Flat package (flatpack) construction (featuring both glass-metal and glass-ceramic seals and welded construction) is popular for planar mounting on flat circuit boards. No holes are required to accommodate the leads of such devices which are arranged on a 0.05 inch pitch (i.e. half the pitch used with DIL devices). Single-in-line (SIL), and quad-in-line (QIL) packages are also becoming increasingly popular whilst TO-5, TO-72, TO-3 and TO-220 encapsulations are also found (the latter being commonly used for three-terminal voltage regulators).

Integrated circuit coding

The European system for coding integrated circuits involves three letters followed by a three or four figure serial number. An additional letter can also be appended to indicate the type of package. The meaning of the letters is as follows:

First letter – type of integrated circuit device:
S digital
T analogue
U mixed analogue/digital
Second letter – series (H indicates hybrid circuit)
Third letter – operational temperature range
A undefined*
B 0°C to +70°C
C −55°C to +125°C
D −25°C to +70°C
E −25°C to +85°C
F −40°C to +85°C
Final letter – package:
B dual-in-line (DIL)
C cylindrical
D dual-in-line (DIL)
F flatpack
P dual-in-line plastic (DIP)
Q quad-in-line
U unpackaged

(* usual for less-critical applications such as consumer electronic equipment)

Many other systems of integrated circuit coding are employed, some of which are peculiar to a particualr manufacturer. Complete device coding often includes the manufacturer's prefix letters.

Example 6.1
Identify each of the following integrated circuits:
(i) *TBA810P*
(ii) *SAA1056P*
(iii) *SAF2039P*

Integrated circuit (i) is an analogue consumer integrated circuit device supplied in a DIP package. The operating temperature range of the device has not been defined. Integrated circuit (ii) is a digital integrated circuit supplied in a DIP package. The operating temperature range of the device has not been defined. Integrated circuit (iii) is a digital consumer integrated circuit device supplied in a DIP package. The operating temperature range for the device is −40°C to +85°C.

Manufacturers' prefix letters for integrated circuits

Prefix	*Manufacturer*	*Prefix*	*Manufacturer*
AD	Analog Devices	PCF	Signetics
Am	Advanced Micro Devices	PIC	Plessey
AH	National Semiconductor	PNA	Mullard
AY	General Instrument	R	Rockwell
C	Intel	R	Raytheon
CD	RCA	RAY	Raytheon
CDP	RCA	RC	Raytheon
D	Intel	S	American Microsystems
DG	Siliconix	SAA	Mullard
DM	National Semiconductor	SAA	Signetics
DS	National Semiconductor	SAB	Mullard
DP	Advanced Micro Devices	SAF	Mullard
DP	National Semiconductor	SCB	Signetics
EF	Thomson/EFCIS	SCN	Signetics
F	Fairchild	SCP	Solid State Scientific
F	Ferranti	SE	Signetics
G	GTE	SL	Plessey
H	SGS	SN	Texas Instruments
HC	Hughes	SP	Plessey
HD	Hitachi	SY	Synertek
HEF	Mullard	TAB	Plessey
HEF	Signetics	TBP	Texas Instruments
HM	Hitachi	TC	Toshiba
HN	Hitachi	TCA	Signetics
I	Intel	TCM	Texas Instruments
ICL	Intersil	TDA	Mullard
IM	Intersil	TDA	Signetics
INS	National Semiconductor	TEA	Signetics
KMM	Texas Instruments	TIB	Texas Instruments
LF	National Semiconductor	TIC	Texas Instruments
LM	National Semiconductor	TIF	Texas Instruments
LM	Signetics	TIL	Texas Instruments
LM	Texas Instruments	TIM	Texas Instruments
LS	Texas Instruments	TIP	Texas Instruments
NM	National Semiconductor	TL	Texas Instruments
M	Mitsubishu	TLC	Texas Instruments
MAB	Mullard	TMM	Toshiba
MBL	Fujitsu	TMP	Texas Instruments
MC	Motorola	TMS	Texas Instruments
MC	Signetics	UA	Signetics
MC	Texas Instruments	UA	Texas Instruments
MJ	Plessey	UCN	Sprague
MK	Mostek	UDN	Sprague

Prefix	*Manufacturer*	*Prefix*	*Manufacturer*
ML	Plessey	UDN	Texas Instruments
MM	National Semiconductor	UGN	Sprague
MN	Plessey	ULN	Signetics
MP	Micro Power Systems	ULN	Sprague
MSM	OKI	ULN	Texas Instruments
MV	Plessey	UPB	NEC
N	Signetics	UPD	NEC
NE	Signetics	VM	Texas Instruments
NJ	Plessey	X	Xicor
NS	National Semiconductor	XR	Raytheon
NSC	National Semiconductor	Z	Zilog
P	Advanced Micro Devices	Z	SGS
P	Intel	ZN	Ferranti
PC	Signetics	μA	Texas Instruments
PCB	Mullard	μPD	NEC
PCF	Mullard		

Digital integrated circuits

Digital integrated circuits have numerous applications quite apart from their obvious use in computing. Digital signals exist only in discrete steps or levels; intermediate states are disallowed. Conventional electronic logic is based on two binary states, commonly referred to as logic 0 (low) and logic 1 (high).

The relative scale of integration of a digital integrated circuit is usually specified in terms of the number of individual semiconductor devices which it contains. The following terminology is commonly used:

Scale of integration	*Abbreviation*	*Number of logic gates**
Small	SSI	1 to 10
Medium	MSI	10 to 100
Large	LSI	100 to 1000
Very large	VLSI	1000 to 10000
Super large	SLSI	10000 to 100000

*or circuitry of equivalent complexity.

Logic gates

The British Standard (BS) and American Standard (MIL/ANSI) symbols for some basic logic gates are shown, together with their *truth tables* in Figure 6.2. The action of each of the basic logic gates is summarised below. Note that, whilst inverters and buffers each have only one input, exclusive-OR gates have two inputs and the other basic gates (AND, OR, NAND and NOR) are commonly available with up to eight inputs.

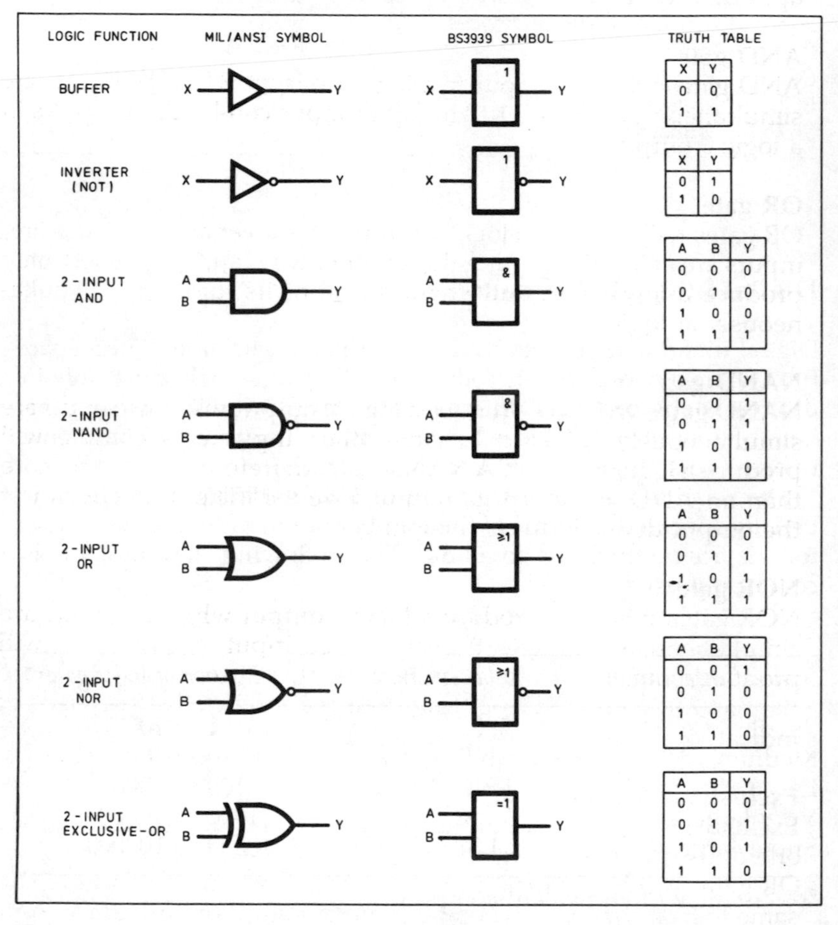

Figure 6.2 Logic gate symbols and truth tables

Buffer
Buffers do not affect the logical state of a digital signal (i.e. a logic 1 input results in a logic 1 output whereas a logic 0 input results in a logic 0 output. Buffers are normally used to provide extra current drive at the output but can also be used to regularise the logic levels present at an interface.

Inverter
Inverters are used to complement the logical state (i.e. a logic 1 input results in a logic 0 output and vice versa). Inverters also provide extra current drive and, like buffers, are used in interfacing applications.

AND gate
AND gates will only produce a logic 1 output when all inputs are simultaneously at logic 1. Any other input combination results in a logic 0 output.

OR gate
OR gates will produce a logic 1 output whenever any one, or more, inputs are at logic 1. Putting this another way, an OR gate will only produce a logic 0 output whenever all of its inputs are simultaneously at logic 0.

NAND gate
NAND gates will only produce a logic 0 output when all inputs are simultaneously at logic 1. Any other input combination will produce a logic 1 output. A NAND gate, therefore, is nothing more than an AND gate with its output inverted! The circle shown at the output denotes this inversion.

NOR gate
NOR gates will only produce a logic 1 output when all inputs are simultaneously at logic 0. Any other input combination will produce a logic 0 output. A NOR gate, therefore, is nothing more than an OR gate with its output inverted. A circle is again used to indicate inversion.

Exclusive-OR gate
Exclusive-OR gates will produce a logic 1 output whenever either one of its inputs is at logic 1 and the other is at logic 0. Exclusive-OR gates produce a logic 0 output whenever both inputs have the same logical state (i.e. when both are at logic 0 or both are at logic 1).

Monostable

A logic device which has only one stable output state is known as a monostable. The output of such a device is initially at logic 0 (low) until an appropriate level change occurs at its trigger input. This level change can be from 0 to 1 (positive edge trigger) or 1 to 0 (negative edge trigger) depending upon the particular monostable device or configuration. Upon receipt of a valid trigger pulse the output of the monostable changes state to logic 1. Then, after a time interval determined by external C-R timing components, the output reverts to logic 0. The device then awaits the arrival of the next trigger.

A common example of a TTL monostable device is the 74121. This device can be triggered by either positive or negative edges depending upon the configuration employed. The chip has complementary inputs (labelled Q and $\overline{Q}$) and requires only two timing components (one resistor and one capacitor).

The internal arrangement of the 74121 is depicted in Figure 6.3. Control inputs A1, A2, and B are used to determine the trigger

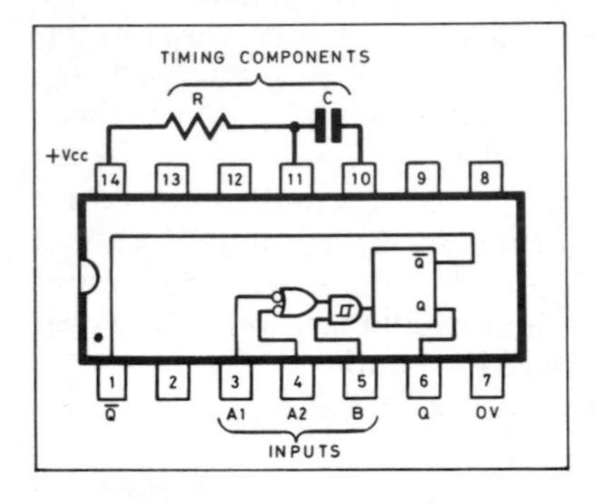

Figure 6.3 Internal arrangement of the 74121 monostable

mode and may be connected in any one of the following three ways:

(a) A1 and A2 connected to logic 0. The monostable will then trigger on a negative edge applied to B.
(b) A1 and B connected to logic 1. The monostable will then trigger on a negative edge applied to A2.
(c) A2 and B connected to logic 1. The monostable will then trigger on a negative edge applied to A1.

Hints and tips

★ Unlike some other astable types, the 74121 is not re-triggerable during its monostable timing period. This simply means that, once a timing period has been started no further trigger pulse will be recognised. Furthermore, in normal use, a recovery time equal in length to the monostable pulse should be allowed before attempting to re-trigger the device.

★ A typical application for a monostable device is in stretching a pulse of very short duration. A 74121 is an ideal device to perform this function; it can be triggered by a very short duration pulse and will continue with its fixed duration timing period long after the input signal has reverted to its original state. The only requirement is that, to ensure reliable triggering, the input pulse should have a width of at least 50ns. For a 74121, the values of external timing resistor should normally lie in the range 1.5kΩ to 47kΩ. The minimum recommended value of external capacitor is only limited by the leakage current of the capacitor employed (in practice values of several hundred microfarads can be used).

Bistable

The output of a bistable has two stable states (logic 0 or logic 1) and, once set, the output of the device will remain at a particular logic level for an indefinite period until reset. Various forms of bistable are available.

R-S bistables

The simplest form of bistable is the R-S bistable (Figure 6.4). This device has two inputs, SET and RESET, and complementary outputs, Q and $\bar{Q}$. A logic 1 applied to the SET input will cause

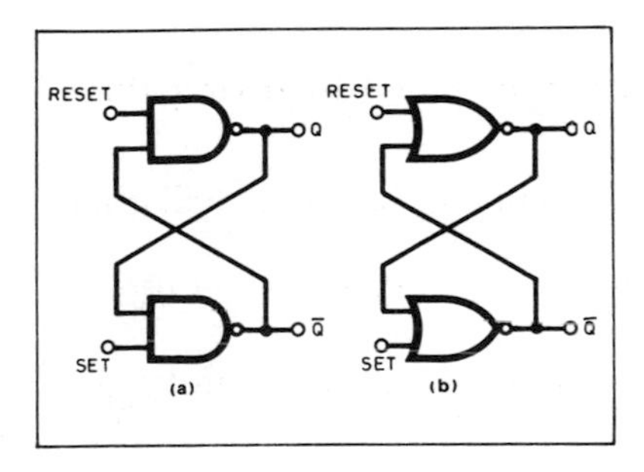

Figure 6.4 Simple R-S bistables based on logic gates: (a) using NAND gates; (b) using NOR gates

the Q output to become (or remain at) logic 1 whilst a logic 1 applied to the RESET input will cause the Q output to become (or remain at) logic 0. In either case, the bistable will remain in its SET or RESET state until an input is applied in such a sense as to change the state.

D-type bistables

The D-type bistable has two principal inputs; D (standing variously for data or delay) and CLOCK (CK). The data input (logic 0 or logic 1) is clocked into the bistable such that the output state only changes when the clock changes state. Operation is thus said to be synchronous. Additional subsidiary inputs (which are invariably active low) are provided which can be used to directly set or reset the bistable. These are usually called PRESET (PR) and CLEAR (CLR). D-type bistables are used both as latches (a simple form of memory) and as binary dividers.

J-K bistables

J-K bistables have two clocked inputs (J and K), two direct inputs (PRESET and CLEAR), a CLOCK (CK) input, and outputs (Q and $\bar{Q}$). As with R-S bistables, the two outputs are complementary (i.e. when one is 0 the other is 1, and vice versa). Similarly, the PRESET and CLEAR inputs are invariably both active low (i.e. a 0 on the PRESET input will set the Q output to 1 whereas a 0 on the CLEAR input will set the Q output to 0).

Hints and tips

★ R-S bistables can be easily implemented using cross-coupled NAND or NOR gates, as shown in Figure 6.4. These arrangements are, however, unreliable as the output state is indeterminate when S and R are simultaneously at logic 1.

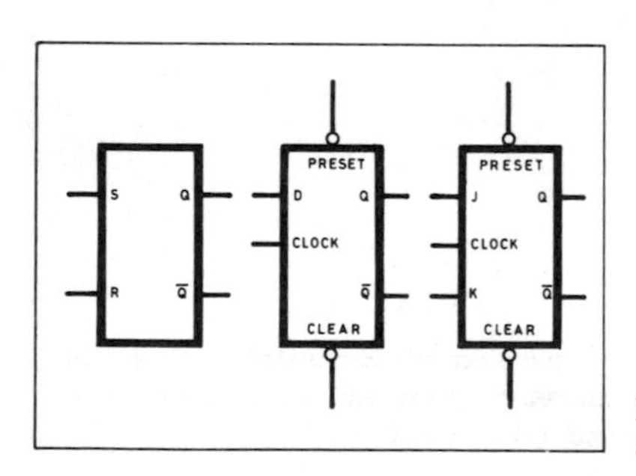

Figure 6.5 Bistable symbols (R-S, D-type and J-K)

★ J-K bistables are the most sophisticated and flexible of the bistable types and they can be configured in various ways including binary dividers, shift registers, and latches.

Logic families

Digital integrated circuit devices are often classified according to the semiconductor technology used in their fabrication; the logic family to which a device belongs being largely instrumental in determining its operational characteristics (such as power consumption, speed, and immunity to noise).

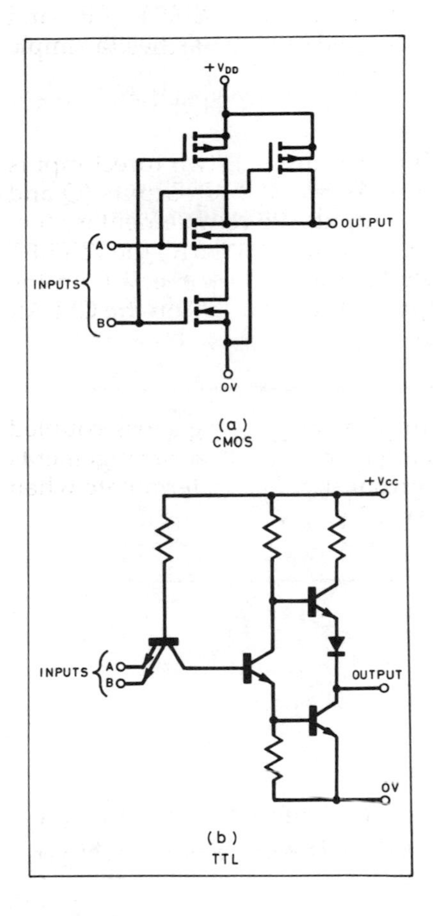

Figure 6.6 Representative two-input AND gate circuits: (a) using CMOS technology; (b) using TTL technology

The two basic logic families are *CMOS* (complementary metal oxide semiconductor) and *TTL* (transistor transistor logic). Each of these families is then further sub-divided. Representative circuits of a two-input AND gate in both technologies are shown in Figure 6.6.

The most common family of TTL logic devices is known as the 74-series. Devices from this family are coded with the prefix number 74. Variants within the family are identified by letters which follow the initial 74 prefix, as follows:

Infix	*Meaning*
none	standard TTL device
C	CMOS version of a TTL device
F	'fast' – a high speed version of the device.
H	high speed version.
S	Schottky input configuration (improve speed and noise immunity).
LS	low power Schottky device
HC	high speed CMOS version (CMOS compatible inputs)
HCT	high speed CMOS version (TTL compatible inputs)

The most common family of CMOS devices is known as the 4000-series. Variants within the family are identified by suffix letters as follows:

Suffix	*Meaning*
none	standard CMOS device
A	standard (unbuffered) CMOS device
B, BE	improved (buffered) CMOS device
UB, UBE	improved (unbuffered) CMOS device

Example 6.2
Identify each of the following integrated circuits:
(i) 4001UBE
(ii) 74LS14

Integrated circuit (i) is an improved (unbuffered) version of the CMOS 4001 device. Integrated circuit (ii) is a low-power Schottky version of the TTL 7414 device.

Logic circuit characteristics

Logic levels are simply the range of voltages used to represent the logic states 0 and 1. The logic levels for CMOS differ markedly from those associated with TTL. In particular, CMOS logic levels are *relative* to the supply voltage used whilst the logic levels associated with TTL devices tend to be *absolute*. The following table usually applies:

	CMOS	*TTL*
logic 1	more than 2/3 V_{DD}	more than 2V
logic 0	less than 1/3 V_{DD}	less than 0.8V
indeterminate	between 1/3 V_{DD} and 2/3 V_{DD}	between 0.8V and 2V

(Note: V_{DD} is the positive supply associated with CMOS devices)

The *noise margin* of a logic device is a measure of its ability to reject noise; the larger the noise margin the better is its ability to perform in an environment in which noise is present. Noise margin is defined as the difference between the minimum values of high state output and high state input voltage and the maximum values of low state output and low state input voltage. Hence:

$$\text{Noise margin} = V_{OH(MIN)} - V_{IH(MIN)}$$
$$\text{or} \quad \text{noise margin} = V_{OL(MAX)} - V_{IL(MAX)}$$

where $V_{OH(MIN)}$ is the minimum value of high state (logic 1) output voltage, $V_{IH(MIN)}$ is the minimum value of high state (logic 1) input

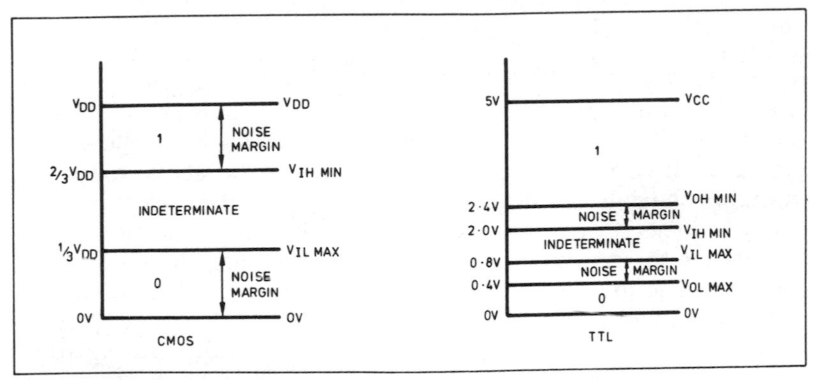

Figure 6.7 Logic levels and noise margins for CMOS and TTL devices

voltage, $V_{OL(MAX)}$ is the maximum value of low state (logic 0) output voltage, and $V_{IL(MIN)}$ is the minimum value of low state (logic 0) input voltage.

The noise margin for standard 7400 series TTL is typically 400mV whilst that for CMOS is 1/3 V_{DD}, as shown in Figure 6.7. The table compares the more important characteristics of various members of the TTL family with buffered CMOS logic:

Characteristic	*Logic family*			
	74	*74LS*	*74HC*	*40BE*
Maximum supply voltage	5.25V	5.25V	5.5V	18V
Minimum supply voltage	4.75V	4.75V	4.5V	3V
Static power dissipation (mW per gate at 100kHz)	10	2	negligible	negligible
Dynamic power dissipation (mW per gate at 100kHz)	10	2	0.2	0.1
Typical propagation delay (ns)	10	10	10	105
Maximum clock frequency (MHz)	35	40	40	12
Speed-power product (pJ at 100kHz)	100	20	1.2	11
Minimum output current (mA at V_O = 0.4V)	16	8	4	1.6
Fan-out (LS loads)	40	20	10	4
Maximum input current (mA at V_I = 0.4V)	−1.6	−0.4	±0.001	−0.001

Microprocessor bus compatible digital integrated circuits invariably have *tri-state* outputs. These outputs can be placed in a high impedance state (i.e. effectively disconnected) in order to avoid bus conflicts which may occur when several logic devices attempt to drive the bus simultaneously. Such devices have a control input called ENABLE (EN) or CHIP SELECT (CS) which allows the device to drive the bus. Such an input may be active high (the output of the gate is valid when the enable or chip select input is taken to logic 1) or may be active low (the output of the gate is valid when the enable or chip select input is taken to logic 0). On the symbol of the device a small circle is often used to denote an active low enable or chip select point.

The *fan-in* of a TTL logic circuit is a measure of the loading effect of its inputs in comparison with a standard TTL gate. A TTL device with a fan-in of two will have inputs which are each equivalent to two standard TTL input loads. The *fan-out* of a logic gate is a measure of its ability to drive further inputs. A TTL device with a fan-out of two will be capable of driving two standard TTL input loads. Clearly, at any node in a digital logic circuit, the fan-out of the driving stage must always be greater than, or equal to, the total fan-in of the following stages.

Driving device	*Maximum number of inputs that may be driven*				
	74	*74LS*	*74S*	*74HC*	*CMOS*
74	10	40	8	unlimited	unlimited
74 buffers	30	60	24	unlimited	unlimited
74LS	5	20	4	unlimited	unlimited
74LS buffers	15	60	12	unlimited	unlimited
74HC	2	10	2	unlimited	unlimited
74HC buffers	4	15	4	unlimited	unlimited
CMOS	–	1	–	50	50

Hints and tips

★ Most TTL and CMOS logic systems are designed to operate from a single supply voltage rail of nominally +5V. With TTL devices, it is important for this voltage to be very closely regulated. Typical TTL i.c. specifications call for regulation of better than ±5% (i.e. the supply voltage should not fall outside the range 4.75V to 5.25V).

★ When operating at reduced supply voltages (particularly with CMOS devices) it is important to note that the propagation delay (i.e. the time taken for a change of state to appear at the output in response to a change at the input) will be significantly increased. In order to maintain performance at high switching speeds, it is important to use a relatively high value of supply voltage.

★ The absolute maximum supply voltage for TTL devices is normally 7V. If the supply voltage ever exceeds this, any TTL devices connected to the supply rail are liable to self destruct very quickly!

★ CMOS logic offers greater tolerance of supply rail variations and operates from a wider range of supply voltages (typically 3V to 15V) than TTL. Coupled with minimal current demand, this makes CMOS an obvious choice of logic family for use with battery operated (portable) equipment.

★ TTL devices require considerably more supply current than their CMOS equivalents. A typical TTL logic gate requires a supply current of around 8mA; approximately 1000 times that of its CMOS counterpart when operating at a typical switching speed of 10kHz.
★ CMOS devices consume negligible power in the *quiescent* state. Power consumption for a CMOS device is, however, proportional to switching speed and, when this is in excess of several megahertz, power consumption may approach (or even exceed) that of a comparable TTL device.
★ When operating from supply voltages in excess of 5V or so, CMOS devices are more immune to noise than their TTL counterparts. This makes CMOS the obvious choice for any application in which noise is likely to be a problem (e.g. motor control).
★ All CMOS devices are now fitted with input static protection diodes but these should not be relied upon for protection and appropriate static precautions should be adopted when handling such devices.
★ Unbuffered CMOS devices exhibit smaller propagation delay but slightly lower noise margin than comparable buffered types.
★ Unused TTL inputs should be pulled-up to logic 1 (V_{CC}) using 1kΩ or 2.2kΩ resistors. One resistor will provide a pull-up for up to 25 unused standard gate inputs. Unused CMOS inputs should be connected to V_{DD} or V_{SS} depending upon the logic function.
★ Both CMOS and TTL logic require low-impedance supplies which are adequately decoupled. Supply borne noise (due to transient spikes) can usually be eliminated by placing capacitors of 100nF and 10μF at strategic points distributed around a PCB layout. As a general rule, one disc or plate capacitor (of between 10nF and 100nF suitably rated) should be fitted for every two to four devices, whilst an electrolytic capacitor (of between 4.7μF and 47μF suitably rated) should be fitted for every eight to ten devices. Buffers (both inverting and non-inverting) and line-drivers will normally require additional (individual) decoupling.

Digital integrated circuit data

TTL 7400 family summary

Device	*Function*	*Package*	*Pinout reference*
7400	gate	14	a (see Fig 6.8)
7401	gate	14	b
7403	gate	14	a
7404	inverter	14	d
7405	inverter	14	d
7406	inverter	14	d
7407	buffer	14	e
7408	gate	14	f
7409	gate	14	f
7410	gate	14	g
7411	gate	14	h
7412	gate	14	g
7413	gate	14	i
7414	inverter	14	j
7415	gate	14	h
7416	inverter	14	d
7417	buffer	14	e
7418	gate	14	k
7421	gate	14	l
7422	gate	14	k
7423	gate	16	m
7425	gate	14	n
7426	gate	14	a
7427	gate	14	o
7428	gate	14	c
7430	gate	14	p
7432	gate	14	q
7433	gate	14	c
7437	gate	14	a
7438	gate	14	a
7440	gate	14	l
7470	bistable	14	r
7472	bistable	14	s
7473	bistable	14	t
7474	bistable	14	u

continued overleaf

Device	*Function*	*Package*	*Pinout reference*
7475	bistable	16	v
7476	bistable	16	w
7478	bistable	14	x
7486	gate	14	y
7490	counter	14	
7491	shift register	14	
7492	counter	14	
7493	counter	14	
7494	shift register	16	
7495	shift register	14	
7496	shift register	16	
74100	bistable	24	
74104	bistable	14	z
74105	bistable	14	A
74107	bistable	14	B
74109	bistable	16	C
74110	bistable	14	D
74111	bistable	16	E
74112	bistable	16	F
74113	bistable	14	G
74114	bistable	14	H
74121	monostable	14	
74122	monostable	14	
74123	monostable	16	
74124	oscillator	16	
74132	gate	14	a
74133	gate	16	
74134	gate	16	I
74135	gate	16	J
74136	gate	14	y
74137	decoder	16	
74138	decoder	16	K
74139	decoder	16	L
74174	bistable	16	M
74175	bistable	16	N
74176	counter	14	
74177	counter	14	
74178	shift register	14	
74179	shift register	16	
74LS240	bus driver	20	O

Device	*Function*	*Package*	*Pinout reference*
74LS241	bus driver	20	P
74LS242	bus transceiver	14	
74LS243	bus transceiver	14	
74LS244	bus driver	20	Q
74LS245	bus transceiver	20	R
74LS260	gate	14	S
74LS266	gate	14	T
74LS365	bus driver	16	U
74LS366	bus driver	16	V
74LS367	bus driver	16	W
74LS368	bus driver	16	X
74LS640	bus transceiver	20	Y
74LS642	bus transceiver	20	Y
74LS641	bus transceiver	20	Z
74LS645	bus transceiver	20	Z

TTL logic gates

Device	*Function*			*Notes*
7400	quad	2-input	NAND	
7401	quad	2-input	NAND	o.c.
7403	quad	2-input	NAND	
7408	quad	2-input	AND	
7409	quad	2-input	AND	o.c.
7410	triple	3-input	NAND	
7411	triple	3-input	AND	
7412	triple	3-input	NAND	o.c.
7413	dual	4-input	NAND	Sch.
7415	triple	3-input	AND	o.c.
7418	dual	4-input	NAND	
7421	dual	4-input	AND	
7422	dual	4-input	NAND	o.c.
7423	dual	4-input	NOR	str.
7425	dual	4-input	NOR	str.
7426	quad	2-input	NAND	o.c.
7427	triple	3-input	NOR	
7428	quad	2-input	NOR	buf.

continued overleaf

Device	*Function*			*Notes*
7430	single	8-input	NAND	
7432	quad	2-input	OR	
7433	quad	2-input	NOR	o.c. buf.
7437	quad	2-input	NAND	
7438	quad	2-input	NAND	o.c.
7440	dual	4-input	NAND	
7486	quad	2-input	Exclusive-OR	
74132	quad	2-input	NAND	Sch.
74133	single	13-input	NAND	
74134	single	12-input	NAND	tri.
74135	quad	2-input	Exclusive-OR	
74136	quad	2-input	Exclusive-OR	
74260	dual	5-input	NOR	
74266	quad	2-input	Exclusive-OR	o.c.

TTL inverters and buffers

Device	*Function*		*Notes*
7400	hex.	inverter	
7405	hex.	inverter	o.c.
7406	hex.	inverter	o.c. h.v.
7407	hex.	buffer	o.c. h.v.
7414	hex.	inverter	Sch.
7416	hex.	inverter	o.c. h.v.
7417	hex.	buffer	o.c. h.v.

TTL bus-drivers, receivers, and transceivers

Device	*Function*		*Notes*
74LS240	octal	bus driver	Sch. buf. tri.
74LS241	octal	bus driver	Sch. buf. tri.
74LS242	quad	bus transceiver	Sch. buf. tri.
74LS243	quad	bus transceiver	Sch. buf. tri.
74LS244	octal	bus driver	Sch. buf. tri.
74LS245	octal	bus transceiver	Sch. buf. tri.
74LS365	hex.	bus driver	buf. tri.
74LS366	hex.	bus driver	buf. tri.
74LS367	hex.	bus driver	buf. tri.

Device	*Function*		*Notes*
74LS368	hex.	bus driver	buf. tri.
74LS640	octal	bus transceiver	Sch. buf. tri.
74LS642	octal	bus transceiver	Sch. buf. o.c.
74LS641	octal	bus transceiver	Sch. buf. o.c.
74LS645	octal	bus transceiver	Sch. buf. tri.

TTL bistables

Device	*Function*	
7470	single	J-K bistable with preset and clear
7472	single	J-K bistable with preset and clear
7473	dual	J-K bistable with clear
7474	dual	D-type with preset and clear
7475	quad	D-type bistable latch
7476	dual	J-K bistable with preset and clear
7478	dual	J-K bistable with preset and clear
74100	dual	8-bit bistable latch
74104	single	J-K with preset and clear
74105	single	J-K with preset and clear
74107	dual	J-K with clear
74109	dual	J-K with preset and clear
74110	single	J-K with preset and clear
74111	dual	J-K with preset and clear
74112	dual	J-K with preset and clear
74113	dual	J-K bistable with preset
74114	dual	J-K bistable with preset and clear
74174	hex.	D-type with clear
74175	quad	D-type

TTL monostables

Device	*Function*
74121	single
74122	single retriggerable with clear
74123	dual retriggerable with clear

TTL counters

Device	*Function*
7490	divide-by-two and divide-by-five
7492	divide-by-two and divide-by-six
7493	divide-by-two and divide-by-eight
74176	presettable, decade
74177	presettable, binary

TTL shift registers

Device	*Function*
7491	8-bit serial-in, serial-out
7494	4-bit, dual asynchronous presets
7495	4-bit, shift-left or shift-right
7496	5-bit, asynchronous preset
74178	4-bit universal
74179	4-bit universal

TTL decoders

Device	*Function*
74LS137	single 3-to-8-line
74LS138	single 3-to-8-line
74LS139	dual 2-to-4-line

TTL shift registers

Device	*Function*
7491	8-bit, serial-in, serial-out
7494	4-bit, dual asynchronous presets
7495	4-bit, shift-left or shift-right
7496	5-bit, asynchronous preset
74178	4-bit universal
74179	4-bit universal

CMOS 4000 family summary

Device	*Function*	*Package*	*Pinout reference*
4000	gate	14	1
4001	gate	14	2
4002	gate	14	3
4006	shift register	14	
4007	transistor array	14	
4008	adder	16	
4009	inverter	16	
4010	buffer	16	
4011	gate	14	4
4012	gate	14	5
4013	bistable	14	
4014	shift register	16	
4015	shift register	16	
4016	analogue switch	14	
4017	counter	16	
4018	counter	16	
4019	gate	16	
4020	counter	16	
4021	shift register	16	
4022	counter	16	
4023	gate	14	6
4024	counter	14	
4025	gate	14	7
4026	display driver	16	
4027	bistable	16	
4028	decoder	16	
4029	counter	16	
4030	gate	14	12
4032	adder	16	
4034	shift register	24	
4035	shift register	16	
4038	adder	16	
4040	counter	16	
4041	inverter/buffer	14	8
4042	bistable	16	
4043	bistable	16	
4044	bistable	16	

continued overleaf

Device	*Function*	*Package*	*Pinout reference*
4045	counter	16	
4046	phase-locked loop	16	
4047	monostable	14	
4048	gate	16	
4049	inverter	16	9
4050	buffer	16	10
4051	analogue mux.	16	
4052	analogue mux.	16	
4053	analogue mux.	16	
4054	decoder/driver	16	
4056	decoder/driver	16	
4060	counter	16	
4066	analogue switch	14	
4067	mux./demux.	24	
4068	gate	14	
4069	inverter	14	11
4070	gate	14	12
4071	gate	14	13
4072	gate	14	14
4073	gate	14	15
4075	gate	14	16
4076	bistable	16	
4077	gate	14	17
4078	gate	14	18
4081	gate	14	19
4082	gate	14	20
4086	gate	14	
4089	binary multiplier	16	
4093	gate	14	21
4094	register	16	
4096	bistable	14	
4097	mux./demux.	24	
4098	monostable	16	
4099	latch	16	
4501	gate	16	
4502	inverter	16	
4503	buffer	16	
4504	level shifter	16	
4506	gate	16	

Device	*Function*	*Package*	*Pinout reference*
4508	latch	24	
4510	counter	16	
4511	decoder/driver	16	
4512	data selector	16	
4513	decoder/driver	18	
4514	decoder	24	
4515	decoder	24	
4516	counter	16	
4517	shift register	16	
4518	counter	16	
4519	multiplexer	16	
4520	counter	16	
4521	divider	16	
4522	divider	16	
4526	divider	16	
4527	multiplier	16	
4528	monostable	16	
4529	data selector	16	
4530	gate	16	
4531	parity tree	16	
4532	encoder	16	
4534	counter	24	
4536	timer	16	
4538	monostable	16	
4539	data selector	16	
4541	timer	14	
4543	decoder/driver	16	
4544	decoder/driver	16	
4547	decoder/driver	16	
4549	register	16	
4551	multiplexer	16	
4553	counter	16	
4555	decoder	16	
4556	decoder	16	
4557	shift register	16	
4558	decoder	16	
4559	register	16	
4560	adder	16	

continued overleaf

Device	*Function*	*Package*	*Pinout reference*
4561	complementer	14	
4562	shift register	14	
4566	timebase generator	16	
4568	counter/comparator	16	
4569	counter	16	
4572	gate	16	
4580	register	24	
4581	ALU	24	
4583	buffer	16	
4584	inverter	16	
4585	comparator	16	
4597	latch	16	
4598	latch	18	
4599	latch	18	
40103	counter	16	
40105	shift register	16	
40106	inverter	14	22
40107	gate	14	
40109	level shifter	16	
40110	counter/driver	16	
40160	counter	16	
40161	counter	16	
40162	counter	16	
40163	counter	16	
40174	bistable	16	
40208	register	24	

CMOS logic gates

Device	*Function*		*Notes*	
4000	dual	2-input	NOR	also single inverter
4001	quad	2-input	NOR	
4002	dual	4-input	NOR	
4011	quad	2-input	NAND	
4012	dual	4-input	NAND	
4019	quad	2-input	AND/OR	
4023	triple	3-input	NAND	
4025	triple	3-input	NOR	
4030	quad	2-input	Exclusive-OR	
4048	single	8-input	multifunction	
4068	single	8-input	NAND	
4070	quad	2-input	Exclusive-OR	
4071	quad	2-input	OR	
4072	dual	4-input	OR	
4073	triple	3-input	AND	
4075	triple	3-input	OR	
4077	quad	2-input	Exclusive-OR	
4078	single	8-input	NOR	
4081	quad	2-input	AND	
4082	dual	4-input	AND	
4086	dual	2-input	AND/OR/invert	
4093	quad	2-input	NAND	
4501	dual	4-input	NAND	also single 2-input OR/NOR
4506	dual	2-input	AND/OR/invert	
4530	dual	5-input	majority gate	
4572	single	2-input	AND	also OR and quad inverter
40107	dual	2-input	NAND	

CMOS buffers and inverters

Device	*Function*		*Notes*
4009	hex.	inverting buffer	
4010	hex.	buffer	
4041	hex.	inverting buffer	
4049	hex.	inverter	
4050	hex.	buffer	
4069	hex.	inverter	
4502	hex.	inverter.	str.
4503	hex.	buffer	tri.
4583	dual	buffer	Sch.
4584	hex.	inverter	Sch.
40106	hex.	inverter	Sch.

CMOS bistables

Device	*Function*		*Notes*
4013	dual	D-type bistable	
4027	dual	J-K bistable	
4042	quad	D-type latch	
4043	quad	R-S latch	tri.
4044	quad	R-S latch	tri.
4076	quad	D-type register	
4096	single	J-K bistable	
40174	hex.	D-type bistable	

CMOS shift registers

Device		*Function*
4006	single	18-bit shift register
4014	single	8-bit shift register
4015	dual	4-bit shift register
4021	single	8-bit shift register
4034	single	8-bit bi-directional
4035	single	4-bit PIPO shift register
4517	dual	64-bit shift register
4557	single	variable length shift register
4562	single	128-bit shift register
40105	single	4-bit × 16 words FIFO register

CMOS counters

Device	*Function*
4017	decade counter
4018	divide-by-N counter
4020	14-bit counter
4022	octal counter
4024	seven-stage ripple counter
4029	presettable binary/BCD up/down counter
4040	12-bit binary counter
4045	21-bit binary counter
4060	14-bit binary counter
4510	BCD up/down counter
4516	binary up/down counter
4518	dual BCD up counter
4520	dual 4-bit binary counter
4534	5-decade counter
4553	3-digit BCD counter
4568	programmable counter/phase comparator
4569	dual programmable BCD counter
40103	8-bit binary presettable synchronous down counter
40110	counter/latch/display driver
40160	asynchronous decade counter with clear
40161	asynchronous 4-bit binary counter with clear
40162	asynchronous decade counter with clear
40163	synchronous 4-bit binary counter with clear

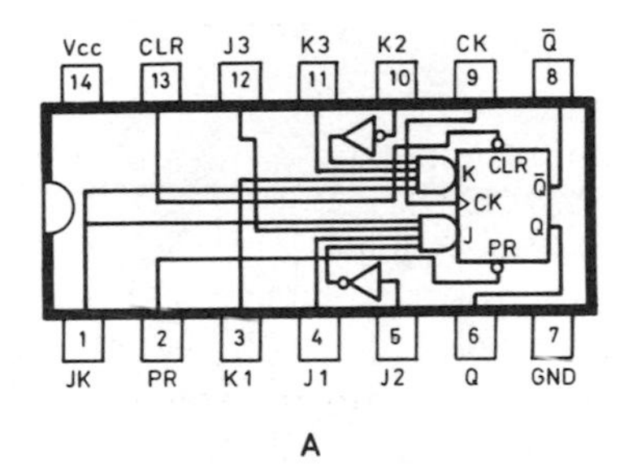

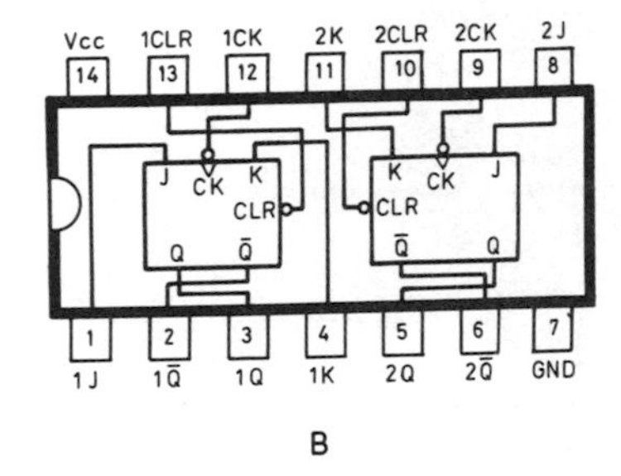

Figure 6.8 Pin connections for common TTL devices

continued overleaf

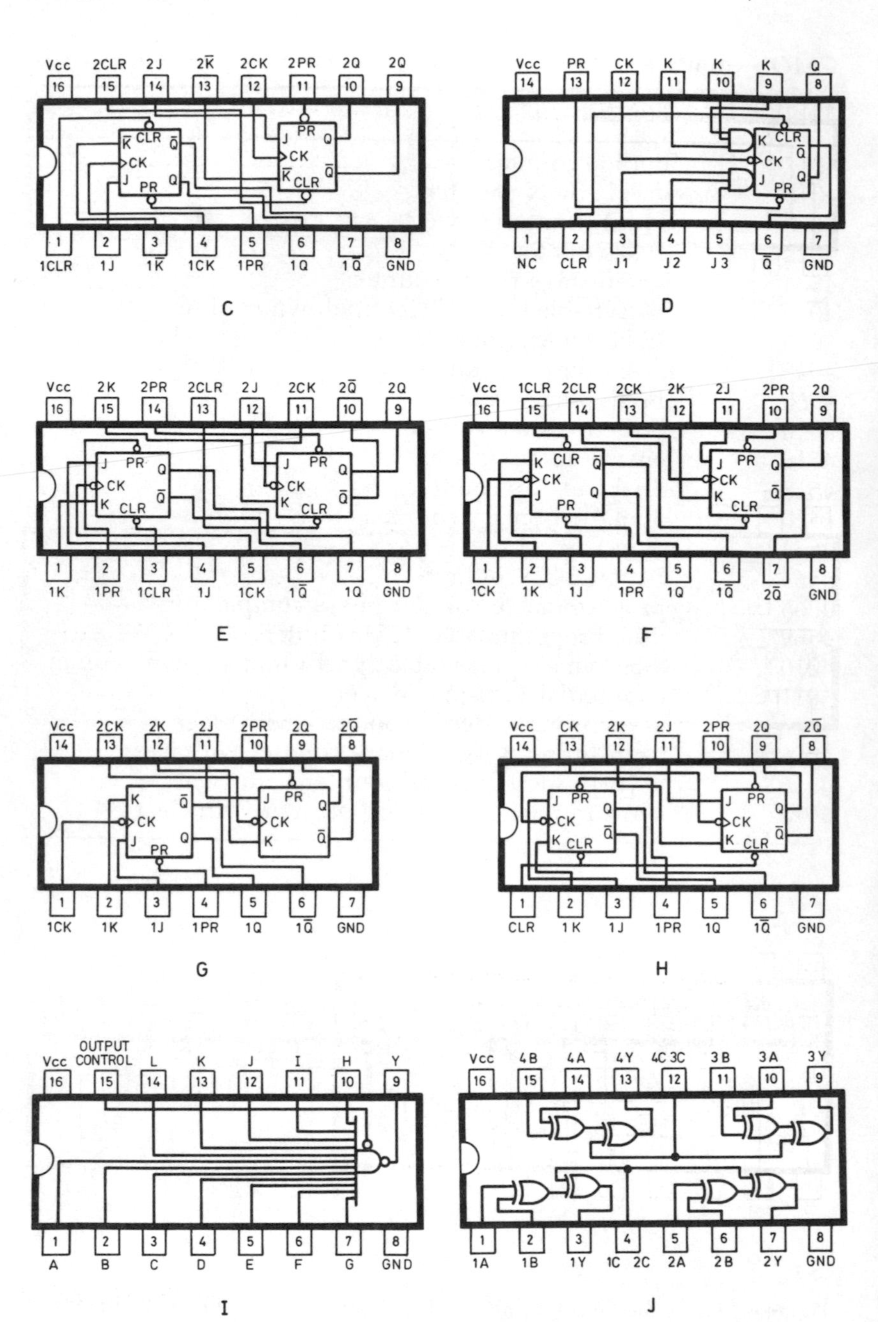
Vcc 2CLR 2J 2K̄ 2CK 2PR 2Q 2Q
16 15 14 13 12 11 10 9
1 2 3 4 5 6 7 8
1CLR 1J 1K̄ 1CK 1PR 1Q 1Q̄ GND
C
Vcc PR CK K K K Q
14 13 12 11 10 9 8
1 2 3 4 5 6 7
NC CLR J1 J2 J3 Q̄ GND
D
Vcc 2K 2PR 2CLR 2J 2CK 2Q̄ 2Q
16 15 14 13 12 11 10 9
1 2 3 4 5 6 7 8
1K 1PR 1CLR 1J 1CK 1Q̄ 1Q GND
E
Vcc 1CLR 2CLR 2CK 2K 2J 2PR 2Q
16 15 14 13 12 11 10 9
1 2 3 4 5 6 7 8
1CK 1K 1J 1PR 1Q 1Q̄ 2Q̄ GND
F
Vcc 2CK 2K 2J 2PR 2Q 2Q̄
14 13 12 11 10 9 8
1 2 3 4 5 6 7
1CK 1K 1J 1PR 1Q 1Q̄ GND
G
Vcc CK 2K 2J 2PR 2Q 2Q̄
14 13 12 11 10 9 8
1 2 3 4 5 6 7
CLR 1K 1J 1PR 1Q 1Q̄ GND
H
Vcc OUTPUT CONTROL L K J I H Y
16 15 14 13 12 11 10 9
1 2 3 4 5 6 7 8
A B C D E F G GND
I
Vcc 4B 4A 4Y 4C 3C 3B 3A 3Y
16 15 14 13 12 11 10 9
1 2 3 4 5 6 7 8
1A 1B 1Y 1C 2C 2A 2B 2Y GND
J

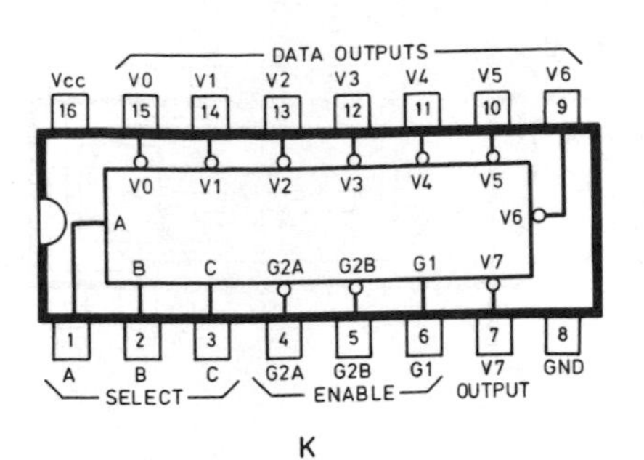

K

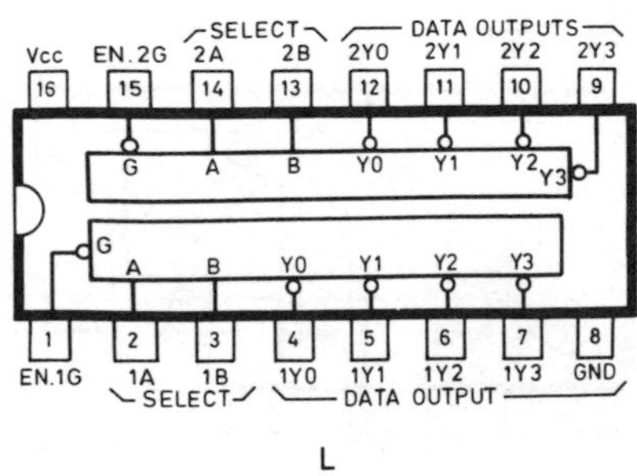

L

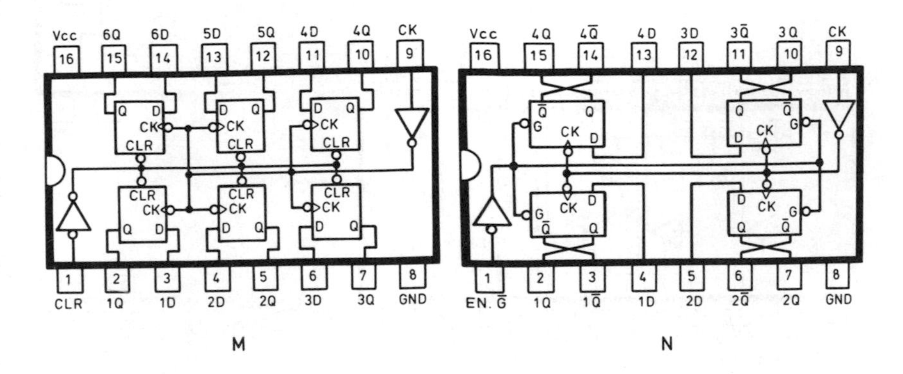

M

N

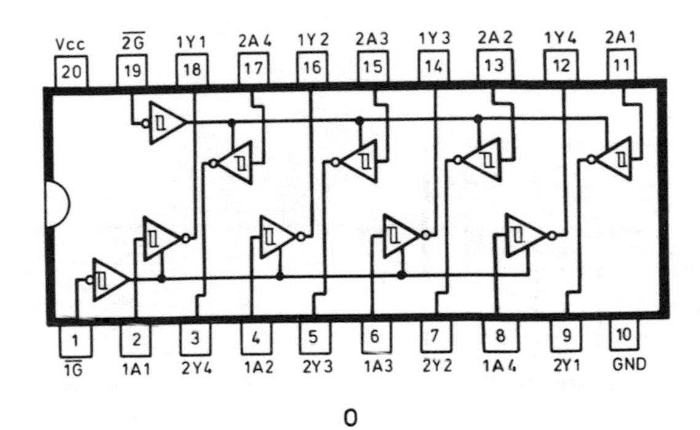

O

continued overleaf

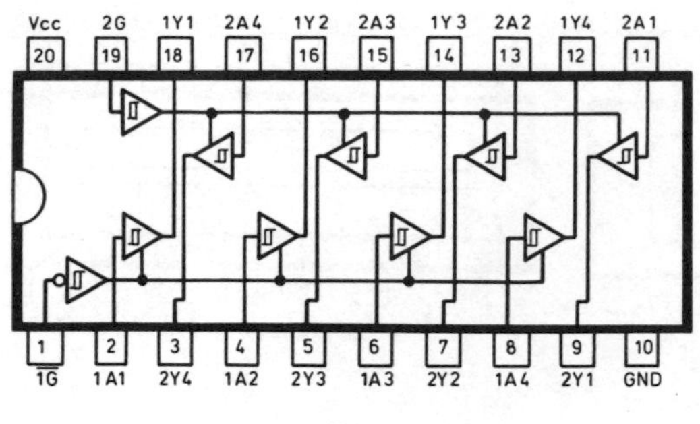
Vcc
2G
1Y1
2A4
1Y2
2A3
1Y3
2A2
1Y4
2A1
20
19
18
17
16
15
14
13
12
11
1
2
3
4
5
6
7
8
9
10
1G
1A1
2Y4
1A2
2Y3
1A3
2Y2
1A4
2Y1
GND

P

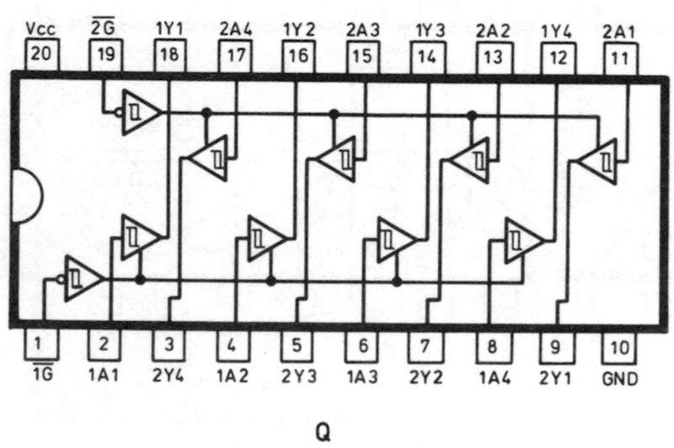
Vcc
2G
1Y1
2A4
1Y2
2A3
1Y3
2A2
1Y4
2A1
20
19
18
17
16
15
14
13
12
11
1
2
3
4
5
6
7
8
9
10
1G
1A1
2Y4
1A2
2Y3
1A3
2Y2
1A4
2Y1
GND

Q

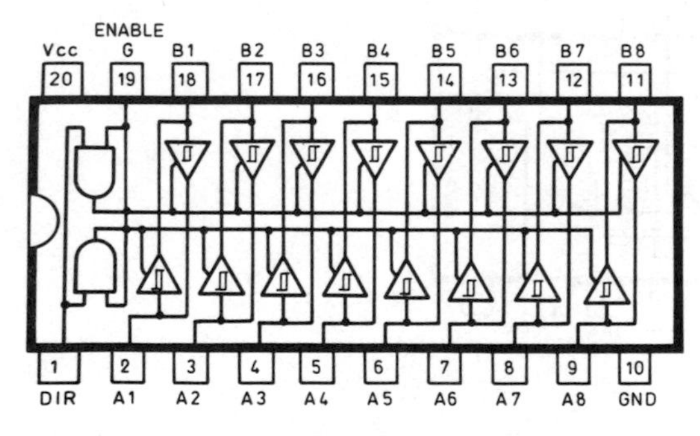
ENABLE
Vcc
G
B1
B2
B3
B4
B5
B6
B7
B8
20
19
18
17
16
15
14
13
12
11
1
2
3
4
5
6
7
8
9
10
DIR
A1
A2
A3
A4
A5
A6
A7
A8
GND

R

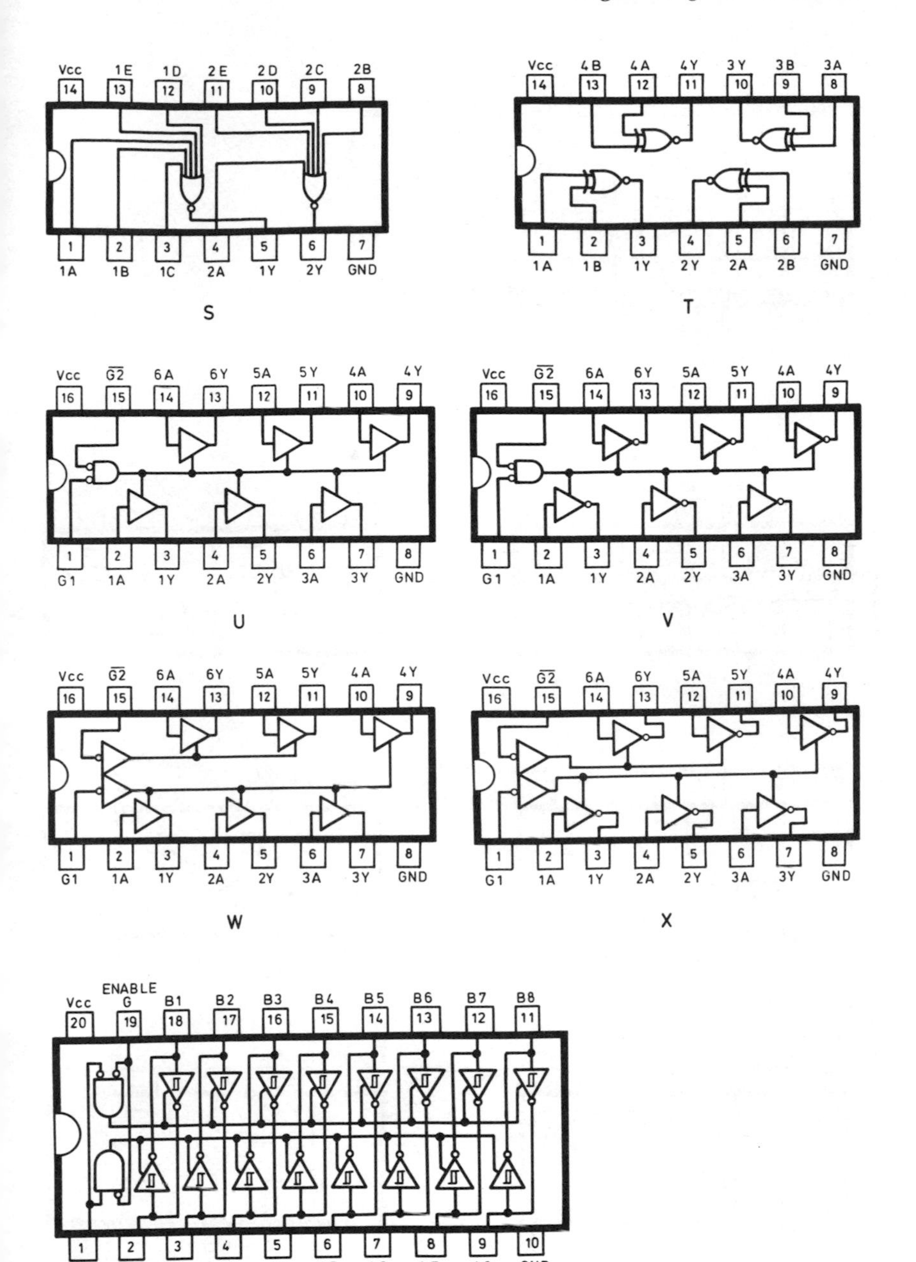

continued overleaf

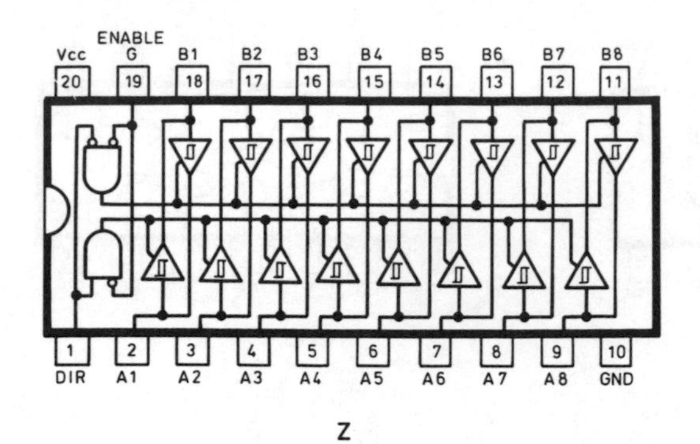
ENABLE
Vcc G B1 B2 B3 B4 B5 B6 B7 B8
20 19 18 17 16 15 14 13 12 11
1 2 3 4 5 6 7 8 9 10
DIR A1 A2 A3 A4 A5 A6 A7 A8 GND

Z

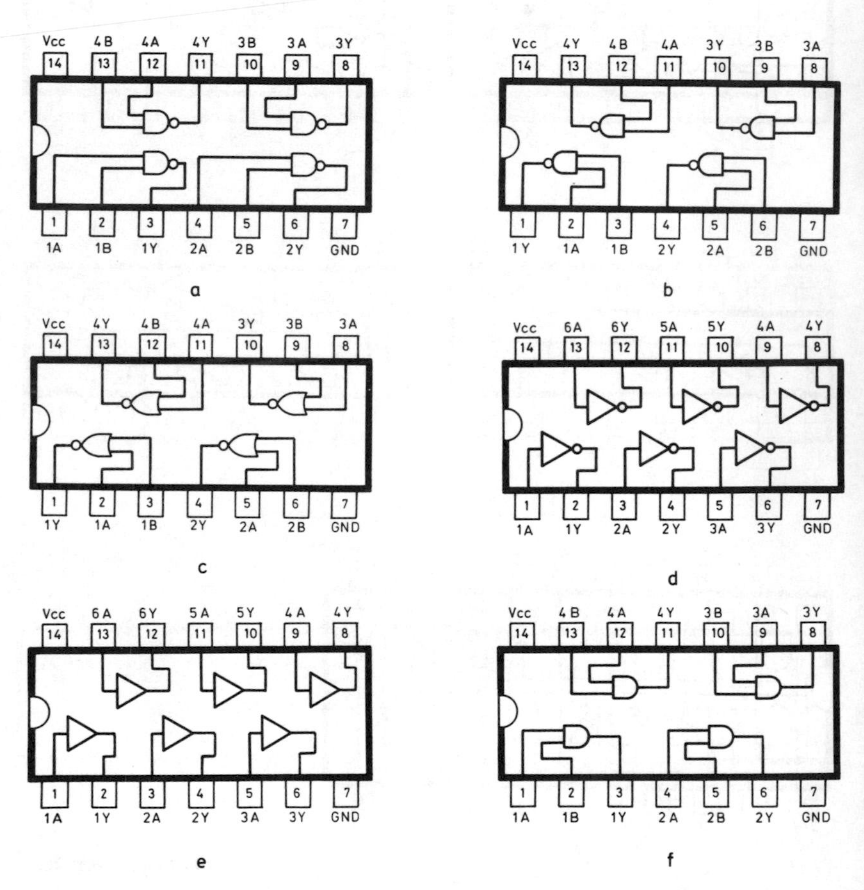
Vcc 4B 4A 4Y 3B 3A 3Y
14 13 12 11 10 9 8
1 2 3 4 5 6 7
1A 1B 1Y 2A 2B 2Y GND
a
Vcc 4Y 4B 4A 3Y 3B 3A
14 13 12 11 10 9 8
1 2 3 4 5 6 7
1Y 1A 1B 2Y 2A 2B GND
b
Vcc 4Y 4B 4A 3Y 3B 3A
14 13 12 11 10 9 8
1 2 3 4 5 6 7
1Y 1A 1B 2Y 2A 2B GND
c
Vcc 6A 6Y 5A 5Y 4A 4Y
14 13 12 11 10 9 8
1 2 3 4 5 6 7
1A 1Y 2A 2Y 3A 3Y GND
d
Vcc 6A 6Y 5A 5Y 4A 4Y
14 13 12 11 10 9 8
1 2 3 4 5 6 7
1A 1Y 2A 2Y 3A 3Y GND
e
Vcc 4B 4A 4Y 3B 3A 3Y
14 13 12 11 10 9 8
1 2 3 4 5 6 7
1A 1B 1Y 2A 2B 2Y GND
f

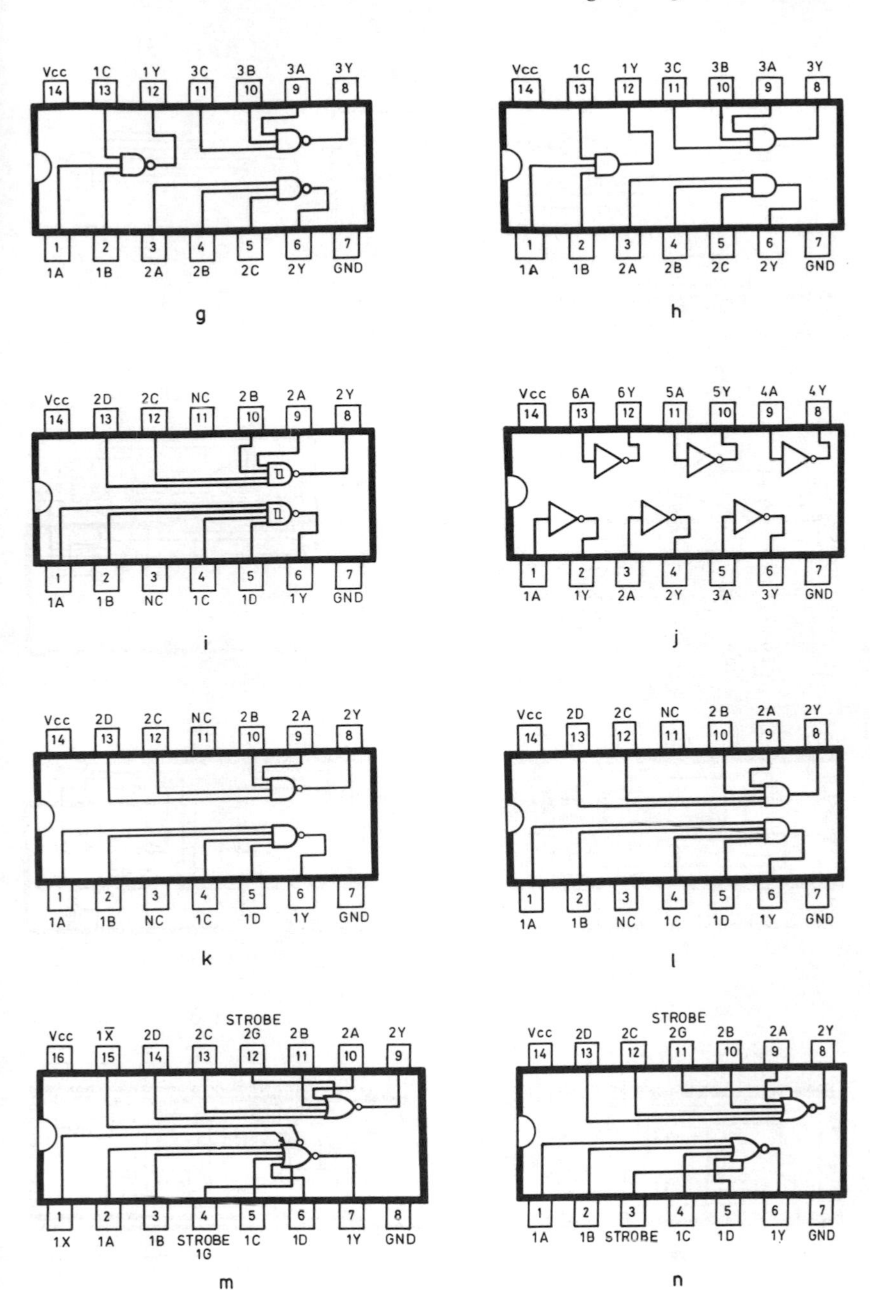

continued overleaf

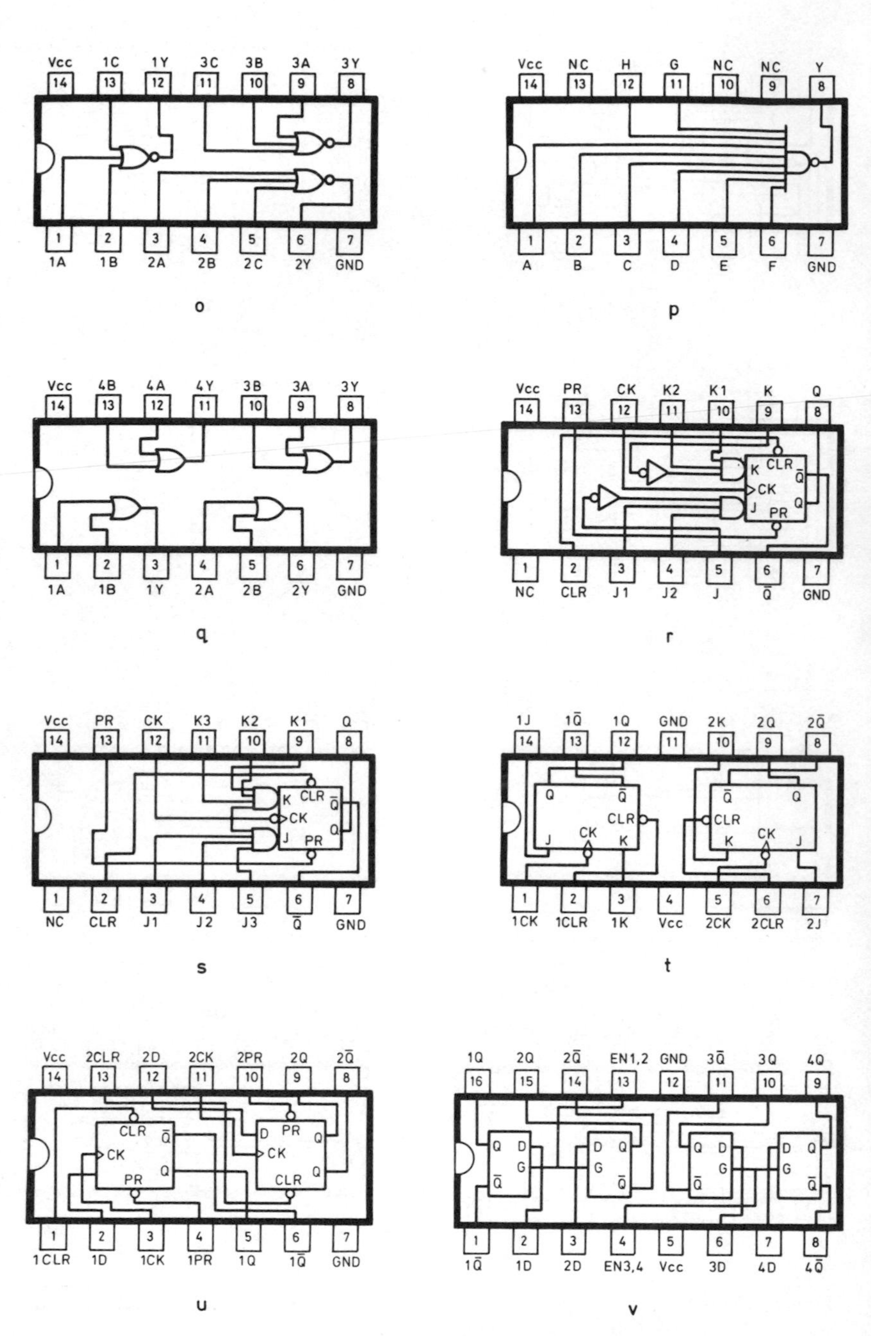
Vcc 1C 1Y 3C 3B 3A 3Y
14 13 12 11 10 9 8
1 2 3 4 5 6 7
1A 1B 2A 2B 2C 2Y GND
o
Vcc NC H G NC NC Y
14 13 12 11 10 9 8
1 2 3 4 5 6 7
A B C D E F GND
p
Vcc 4B 4A 4Y 3B 3A 3Y
14 13 12 11 10 9 8
1 2 3 4 5 6 7
1A 1B 1Y 2A 2B 2Y GND
q
Vcc PR CK K2 K1 K Q
14 13 12 11 10 9 8
CLR K CK J PR Q̄ Q
1 2 3 4 5 6 7
NC CLR J1 J2 J Q̄ GND
r
Vcc PR CK K3 K2 K1 Q
14 13 12 11 10 9 8
K CLR CK J PR Q̄ Q
1 2 3 4 5 6 7
NC CLR J1 J2 J3 Q̄ GND
s
1J 1Q̄ 1Q GND 2K 2Q 2Q̄
14 13 12 11 10 9 8
Q Q̄ CLR J CK K
Q̄ Q CLR K CK J
1 2 3 4 5 6 7
1CK 1CLR 1K Vcc 2CK 2CLR 2J
t
Vcc 2CLR 2D 2CK 2PR 2Q 2Q̄
14 13 12 11 10 9 8
CLR Q̄ CK PR Q
D PR Q CK CLR Q
1 2 3 4 5 6 7
1CLR 1D 1CK 1PR 1Q 1Q̄ GND
u
1Q 2Q 2Q̄ EN1,2 GND 3Q̄ 3Q 4Q
16 15 14 13 12 11 10 9
Q D G Q̄
D Q G Q̄
Q D G Q̄
D Q G Q̄
1 2 3 4 5 6 7 8
1Q̄ 1D 2D EN3,4 Vcc 3D 4D 4Q̄
v

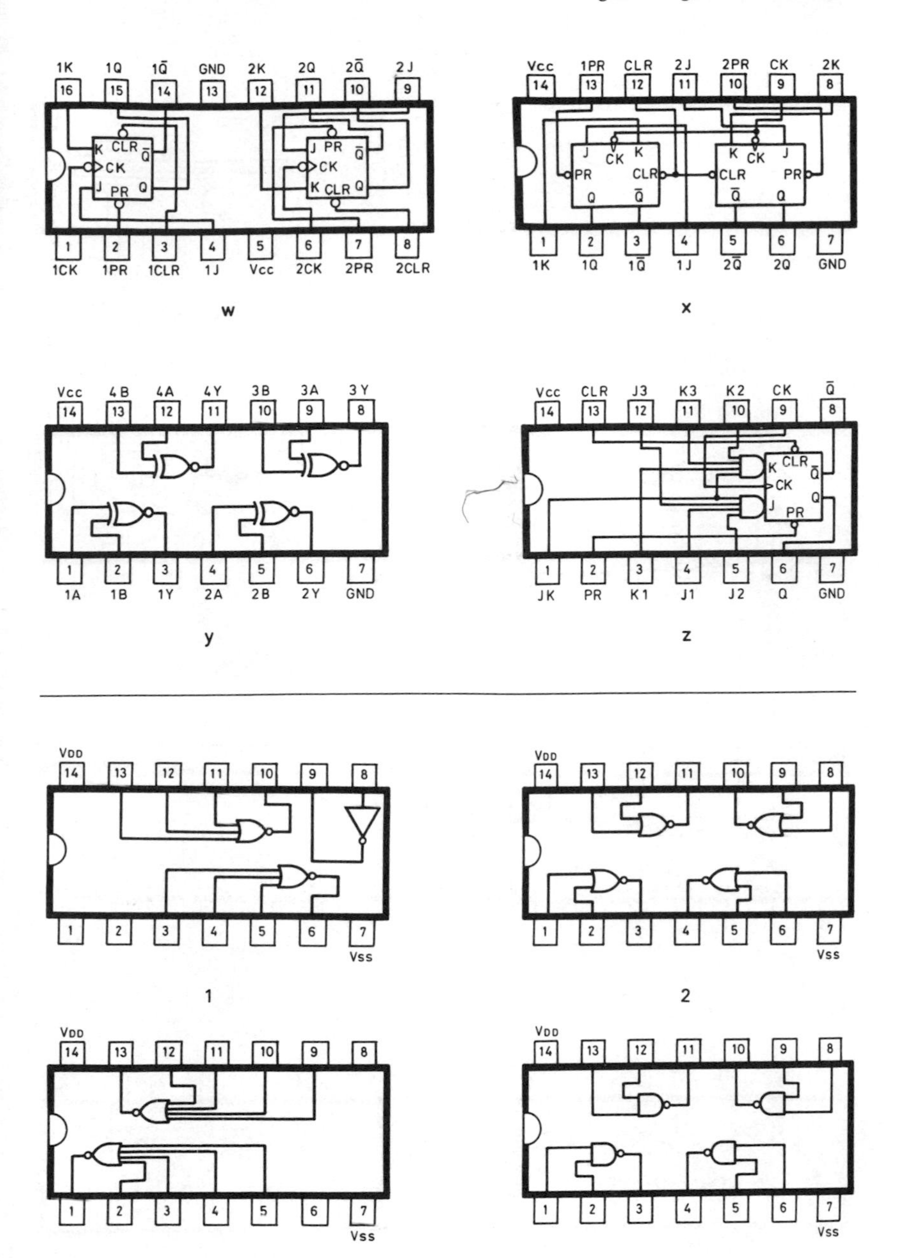

Figure 6.9 Pin connections for common CMOS devices

continued overleaf

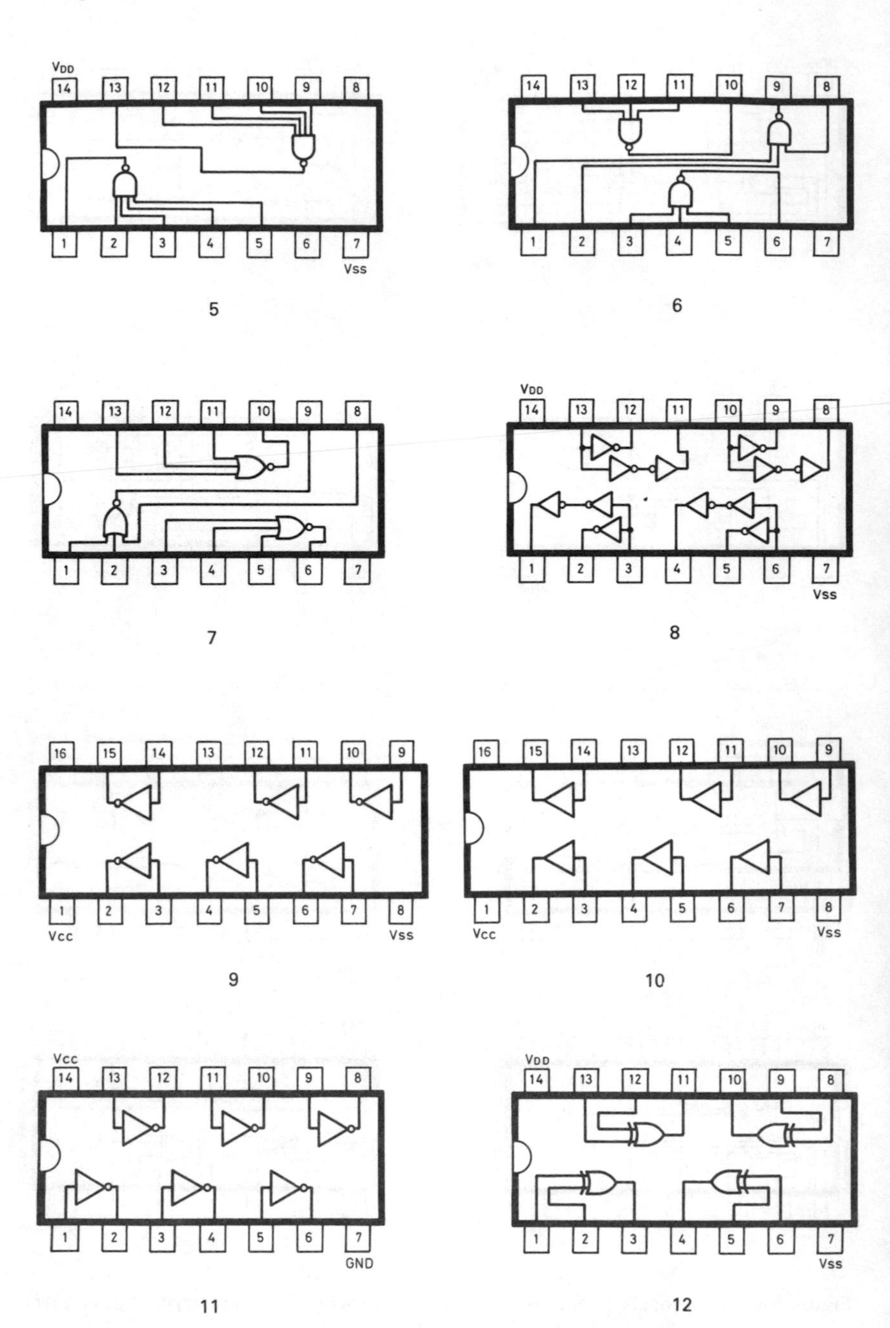
VDD
14
13
12
11
10
9
8
1
2
3
4
5
6
7
Vss
5
14
13
12
11
10
9
8
1
2
3
4
5
6
7
6
14
13
12
11
10
9
8
1
2
3
4
5
6
7
7
VDD
14
13
12
11
10
9
8
1
2
3
4
5
6
7
Vss
8
16
15
14
13
12
11
10
9
1
2
3
4
5
6
7
8
Vcc
Vss
9
16
15
14
13
12
11
10
9
1
2
3
4
5
6
7
8
Vcc
Vss
10
Vcc
14
13
12
11
10
9
8
1
2
3
4
5
6
7
GND
11
VDD
14
13
12
11
10
9
8
1
2
3
4
5
6
7
Vss
12

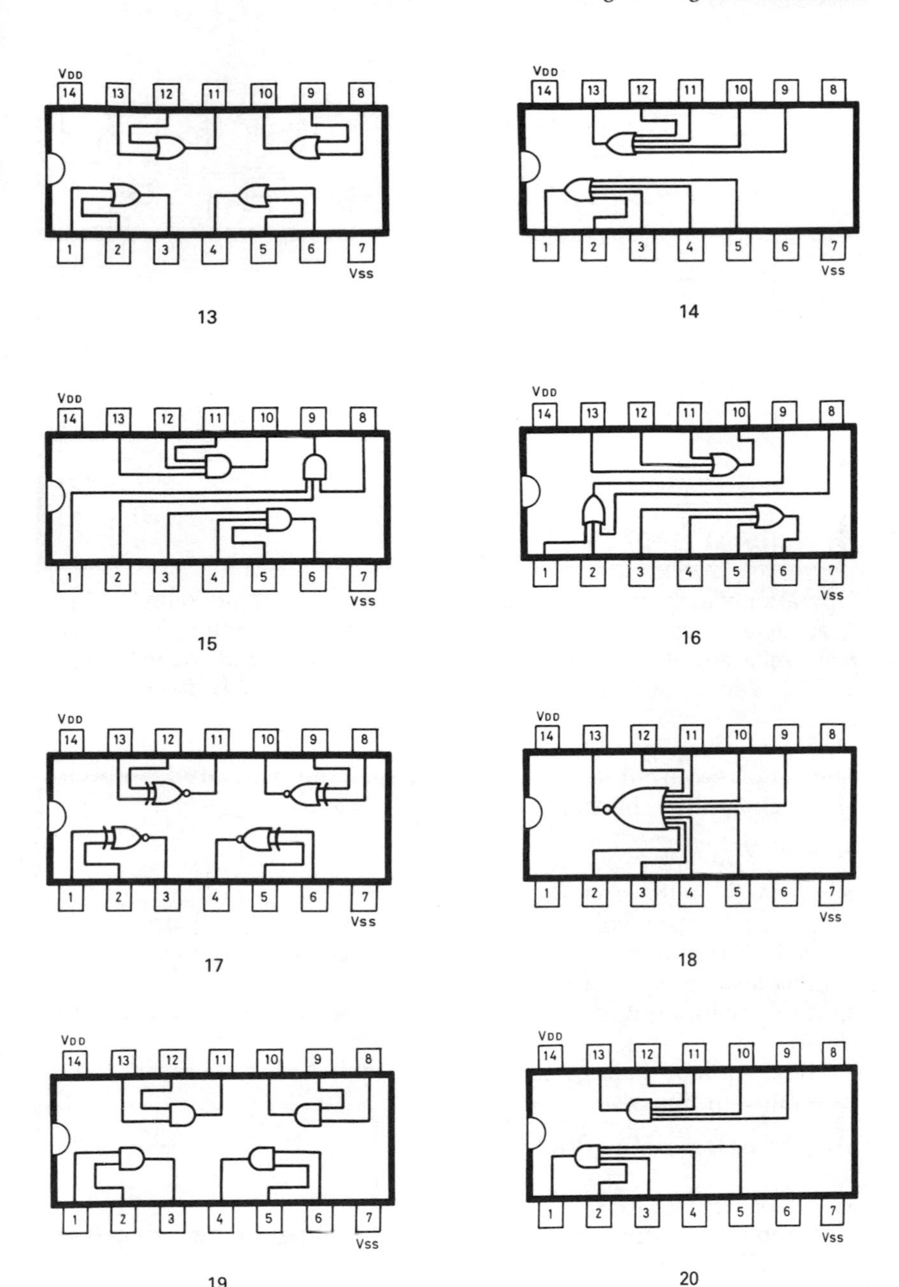

13

14

15

16

17

18

19

20

continued overleaf

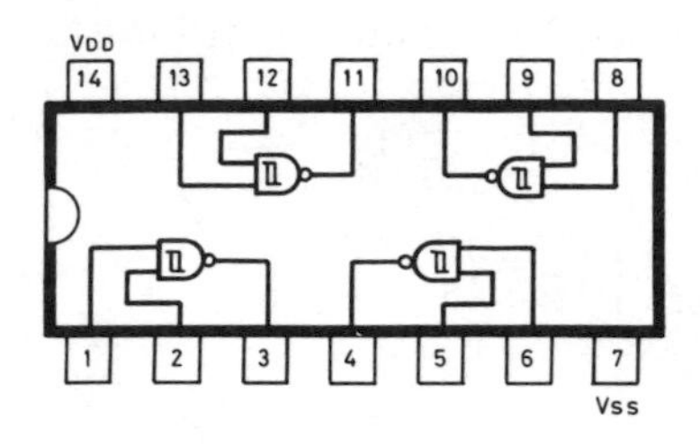

21

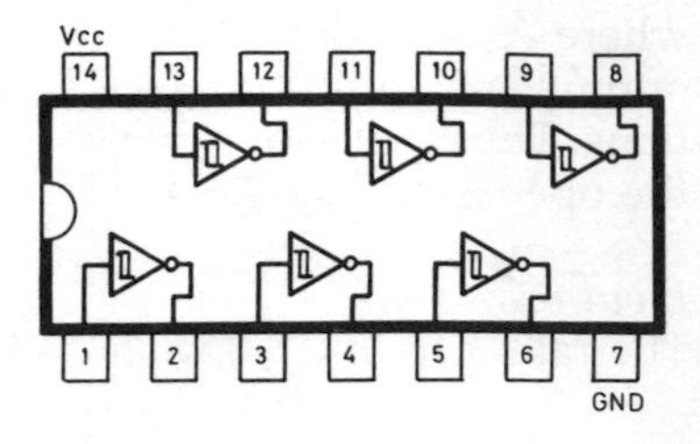

22

Operational amplifiers

Operational amplifiers are analogue integrated circuits which offer near-ideal characteristics (virtually infinite voltage gain and input resistance coupled with low output resistance and wide bandwidth). The following terminology is applied to such devices:

Open-loop voltage gain
The ratio of output voltage to input voltage measured without feedback applied, hence:

$A_{VOL} = V_{OUT}/V_{IN}$

where A_{VOL} is the open-loop voltage gain, V_{OUT} and V_{IN} are the output and input voltages respectively under open-loop conditions. In linear voltage amplifying applications, a large amount of *negative feedback* will normally be applied and the open-loop voltage gain can be thought of as the internal voltage gain provided by the device.

The open-loop voltage gain is often expressed in dB rather than as a ratio. In this case:

$A_{VOL} = 20 \log_{10} (V_{OUT}/V_{IN})$

Closed-loop voltage gain
The ratio of output voltage to input voltage when negative feedback is applied, hence:

$A_{VCL} = V_{OUT}/V_{IN}$

where A_{VCL} is the closed-loop voltage gain, V_{OUT} and V_{IN} are the output and input voltages respectively under closed-loop conditions. The closed-loop voltage gain is normally very much less than the open-loop voltage gain.

Input resistance
The ratio of input voltage to input current:

$$R_{IN} = V_{IN}/I_{IN}$$

where R_{IN} is the input resistance (in Ω), V_{IN} is the input voltage (in V) and I_{IN} is the input current (in A). Note that we usually assume that the input of an operational amplifier is purely resistive though this may not be the case at high frequencies where shunt capacitive reactance may become significant.

The input resistance of operational amplifiers is very much dependent on the semiconductor technology employed. In practice, values range from about 2MΩ for bipolar operational amplifiers to over $10^{12}\Omega$ for CMOS devices.

Output resistance
The ratio of open-circuit output voltage to short-circuit output current, hence:

$$R_{OUT} = V_{OUT(OC)}/I_{OUT(SC)}$$

where R_{OUT} is the output resistance (in Ω), $V_{OUT(OC)}$ is the output voltage (in V) and $I_{OUT(SC)}$ is the output current (in A).

Input offset voltage
This is the voltage which, when applied at the input, provides an output voltage of exactly zero. Similarly, the input offset current is the current which, when applied at the input, provides an output voltage of exactly zero. (Note that, due to imperfect balance and very high internal gain a small output voltage may apper with no input present.) Offset may be minimised by applying large amounts of negative feedback or by using the offset null facility provided by a number of operational amplifiers (see Figure 6.10).

Slew rate
The slew-rate of an operational amplifier is the rate of change of output voltage with time in response to a perfect step-function input. Hence:

$$\text{slew-rate} = \Delta V_{OUT}/\Delta t$$

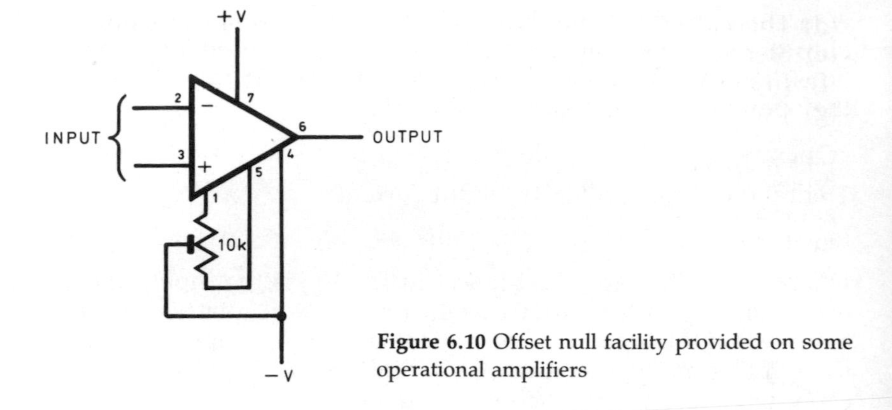

Figure 6.10 Offset null facility provided on some operational amplifiers

where ΔV_{OUT} is the change in output voltage (in V) and Δt is the corresponding internal of time (in s).

Common-mode rejection ratio (CMRR)
The ratio of differential voltage gain to common-mode voltage gain (usually expressed in dB). Hence:

$$CMRR = 20 \log_{10} (A_{VOL(DM)}/A_{VOL(CM)})$$

where $A_{VOL(DM)}$ is the open-loop voltage gain in differential mode (equal to A_{VOL}), and $A_{VOL(CM)}$ is the open-loop voltage gain in common-mode (i.e. signal applied with both inputs connected together). CMRR is a measure of an operational amplifier's ability to reject signals (e.g. noise) which are simultaneously present on both inputs.

Maximum output voltage swing
This is the maximum range of output voltages that the device can produce without distortion. Normally these will be symmetrical about 0V and within a volt, or so, of the supply voltage rails (both positive and negative).

The desirable characteristics for an operational amplifier are as follows:

(a) The open-loop voltage gain should be very high (ideally infinite).
(b) The input resistance should be very high (ideally infinite).
(c) The output resistance should be very low (ideally zero).

(d) The output voltage swing should be as large as possible.
(e) Slew-rate should be as large as possible.
(f) Input offset should be as small as possible.
(g) Common-mode rejection ratio should be as large as possible.

Operational amplifiers are available packaged singly, in pairs (dual types), or in fours (quad types). The 081, for example, is a single general purpose BIFET operational amplifier which is also available in dual (082) and quad (084) forms. Figure 6.11 shows the pin connections used for the most commonly encountered types.

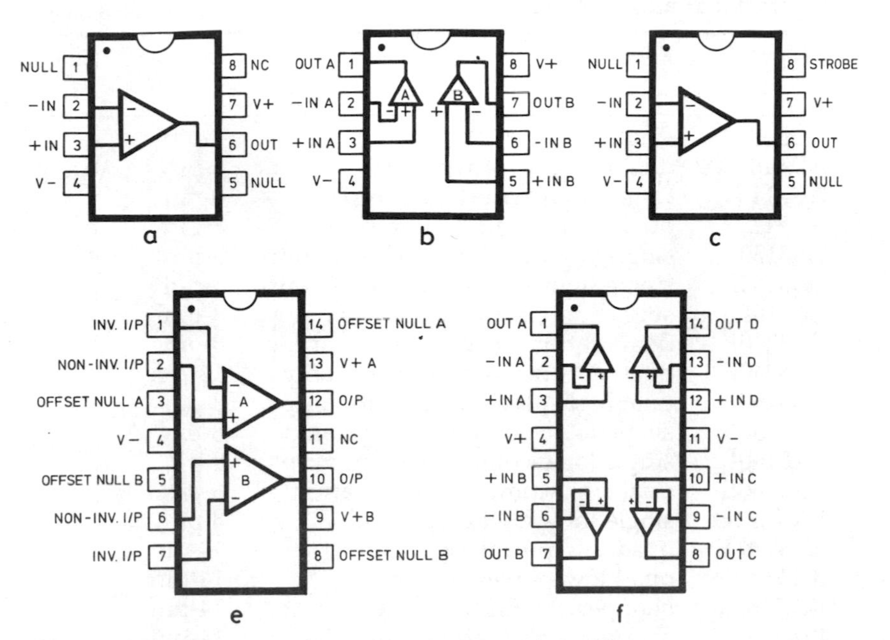

Figure 6.11 Pin connections for common operational amplifiers

Hints and tips

★ Where several operational amplifiers are required in a circuit, it obviously makes good sense to use dual or quad devices rather than single devices. Not only will this minimise the total number of components but it will also reduce the space required on the PCB, improve reliability, and reduce overall cost.

★ Anti-static precautions should be observed when handling CMOS and FET input operational amplifiers.

★ Offset null circuitry can usually be omitted when operational amplifiers are operated in a.c. coupled (capacitor coupled) circuits.
★ The maximum output voltage swing for an operational amplifier is usually quoted under open-circuit output conditions. The undistorted output voltage swing will be somewhat less than that quoted when the output is loaded.

Operational amplifier data

Operational amplifier summary

Device	*Function*	*Package*	*Pinout ref.*
AD548	single low-power op. amp.	8-pin	a
AD648	dual low-power op. amp.	8-pin	b
AD711	single high-speed op. amp.	8-pin	a
AD712	dual high-speed op. amp.	8-pin	b
CA3130E	single high-performance op. amp.	8-pin	c
CA3140E	single high-performance op. amp.	8-pin	c
CA3160E	single high-performance op. amp.	8-pin	
CA3240E	dual high-performance op. amp.	8-pin	b
CA3240E-1	dual high-performance op. amp.	14-pin	e
CA3260E	dual high-performance op. amp.	8-pin	
CA5130E	single logic-compatible op. amp.	8-pin	c
CA5160E	single logic-compatible op. amp.	8-pin	c
CA5260E	dual logic-compatible op. amp.	8-pin	b
ICL7611D	single low-power op. amp.	8-pin	
ICL7641E	quad low-power op. amp.	14-pin	f
ICL7642E	quad low-power op. amp.	14-pin	f
ICL7650S	chopper stabilised op. amp.	14-pin	
ICL7652C	chopper stabilised op. amp.	14-pin	
LF347N	quad high-performance op. amp.	14-pin	f
LG351N	single high-performance op. amp.	8-pin	a
LF353N	dual high-performance op. amp.	8-pin	b
LF355N	single high-performance op. amp.	8-pin	a
LM301N	single general purpose op. amp.	8-pin	
LM308N	single low-drift op. amp.	8-pin	
LM324N	quad general purpose op. amp.	14-pin	f
LM348N	quad general purpose op. amp.	14-pin	f
LM358N	dual logic-compatible op. amp.	8-pin	
NE531N	single high-performance op. amp.	8-pin	

Device	*Function*	*Package*	*Pinout ref.*
NE5532	dual low-noise op. amp.	8-pin	b
NE5534	single low-noise op. amp.	8-pin	
NE5539	single wide bandwidth op. amp.	14-pin	
OP-07CN	single precision op. amp.	8-pin	g
OP-27G	single precision op. amp.	8-pin	g
OP-37GN	single precision op. amp.	8-pin	g
OP-42FZ	single high-performance op. amp.	8-pin	a
OP-77GP	single precision op. amp.	8-pin	g
RC4558P	dual high-performance op. amp.	8-pin	b
TL061CP	single low-power op. amp.	8-pin	a
TL062CP	dual low-power op. amp.	8-pin	b
TL064CN	quad low-power op. amp.	14-pin	f
TL071CP	single low-noise op. amp.	8-pin	a
TL072CP	dual low-noise op. amp.	8-pin	b
TL074CN	quad low-noise op. amp.	14-pin	f
TL081CP	single high-performance op. amp.	8-pin	a
TL082CP	dual high-performance op. amp.	8-pin	b
TL084CN	quad high-performance op. amp.	14-pin	f
741	single general purpose op. amp.	8-pin	a
741S	single high-speed op. amp.	8-pin	a
747	dual general purpose op. amp.	8-pin	e
748	single general purpose op. amp.	8-pin	
759	power operational amplifier	Tab	

Operational amplifier characteristics

Device	*Type*	*Supply voltage range (V)*	*Open loop voltage gain (dB)*	*Input bias current (A)*	*Slew rate (V/µs)*	*Output voltage (V) (note 1)*
AD548	bipolar	±4.5 to ±18	100 min.	0.01n	1.8	±13
AD648	bipolar	±4.5 to ±18	100 min.	0.01n	1.8	±13
AD711	FET	±4.5 to ±18	100	25p	20	±13

continued overleaf

Device	*Type*	*Supply voltage range (V)*	*Open loop voltage gain (dB)*	*Input bias current (A)*	*Slew rate (V/μs)*	*Output voltage (V) (note 1)*
AD712	FET	±4.5 to ±18	100	25p	20	±13
CA3130	CMOS	6 to 16 (or ±3 to ±8)	110	5p	10	13
CA3140	CMOS	4 to 36 (or ±2 to ±18)	100	5p	9	13
CA3160	CMOS	5 to 16	110	5p	10	13.3
CA3240	CMOS	4 to 36 (or ±2 to ±18)	100	5p	9	13
CA3260	CMOS	4 to 16 (or ±2 to ±8)	110	5p	10	13.3
CA5130	CMOS	5 to 16 (or ±2.5 to ±8)	110	5p	10	13.3
CA5160	CMOS	5 to 16 (or ±2 to ±8)	102	2p	10	13.3
CA5260	CMOS	4.5 to 16 (or ±2.25 to ±8)	80	2p	8	4.7 (note 2)
ICL761	CMOS	±9 max.	98	1p	1.6	±4.5 (note 2)
ICL764	CMOS	±9 max.	98	1p	1.6	±4.5 (note 2)
ICL764	CMOS	±9 max.	98	1p	1.6	±4.5 (note 2)
ICL765	CMOS	±4 to ±16	150	1.5p	2.5	±4.85 (note 2)
ICL765	CMOS	±2 to ±8	150	15p	0.5	±4.85 (note 2)
LF347	FET	±5 to ±18	110	50p	13	±13.5
LF351	FET	±5 to ±18	110	50p	13	±13.5
LF353	FET	±5 to ±18	110	50p	13	±13.5
LF355	FET	±4 to ±18	106	30p	5	±13
LM301	bipolar	±5 to ±18	88	70n	0.4	±13

Device	*Type*	*Supply voltage range (V)*	*Open loop voltage gain (dB)*	*Input bias current (A)*	*Slew rate (V/μs)*	*Output voltage (V) (note 1)*
LM308	bipolar	±5 to ±18	102	1.5n		±13
LM324	bipolar	3 to 32 (or ±15)	100	45n		28 (or ±14) (note 3)
LM348	bipolar	±10 to ±18	96	30n	0.6	28 (note 3)
LM358	bipolar	3 to 30	100	40n	0.6	28 (note 3)
NE531	bipolar	±5 to ±22	96	400n	35	±15
NE5532	bipolar	±3 to ±20	100	200n	9	±13
NE5534	bipolar	±3 to ±20	100	500n	13	±13.5
NE5539	bipolar	±8 to ±12	52	5m	600	+2.7 −2.2 (note 4)
OP-07	bipolar	±3 to ±18	132	±2.2n	0.17	±13
OP-27	bipolar	±4 to ±18	123	±15n	2.8	±13
OP-37	bipolar	±4 to ±18	123	±15n	17	±13
OP-42	FET	±20 max.	108	130p	50	+12.5 −11.9
OP-77	bipolar	±3 to ±18	135	1.2n	0.3	±13
RC4558	bipolar	±3 to ±18	85	150n	1.7	±13
TL061	FET	±3.5 to ±18	76	30p	3.5	±13.5
TL062	FET	±3.5 to ±18	76	30p	3.5	±13.5
TL064	FET	±3.5 to ±18	76	30p	3.5	±13.5
TL071	FET	±3 to ±18	106	30p	13	±13.5
TL072	FET	±3 to ±18	106	30p	13	±13.5
TL074	FET	±3 to ±18	106	30p	13	±13.5
TL081	FET	±3 to ±18	106	30p	13	±13.5
TL082	FET	±3 to ±18	106	30p	13	±13.5
TL084	FET	±3 to ±18	106	30p	13	±13.5
741	bipolar	±5 to ±18	106	80n	0.5	±13
741S	bipolar	±5 to ±18	100	200n	20	±13
747	bipolar	±7 to ±18	106	80n	0.5	±13

continued overleaf

Device	*Type*	*Supply voltage range (V)*	*Open loop voltage gain (dB)*	*Input bias current (A)*	*Slew rate (V/μs)*	*Output voltage (V) (note 1)*
748	bipolar	±7 to ±18	106	80n	0.8	±13
759	bipolar	7 to 36 (or ±3.5 to ±18)	106	50n	0.5	±12.5

Notes 1. Measured with supplies of 15V (or ±15V).
2. Measured with supplies of 5V (or ±5V).
3. Measured with supplies of 30V.
4. Measured with supplies of ±8V.

Timers

The generic timer is the 555 device. This versatile integrated circuit is based on a neat hybrid arrangement of analogue and digital

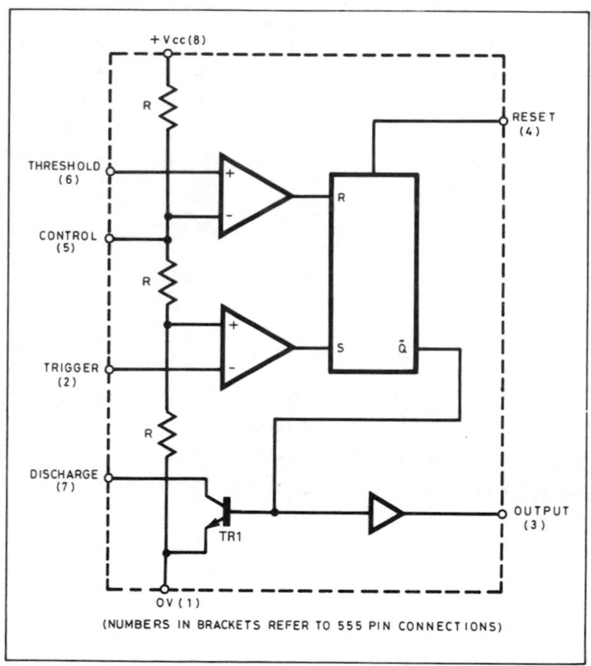

Figure 6.12 Internal arrangement of the 555 timer

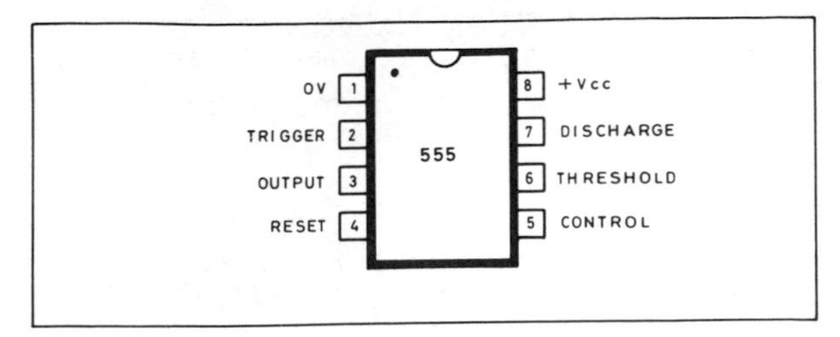

Figure 6.13 Pin connections for the standard 555 timer

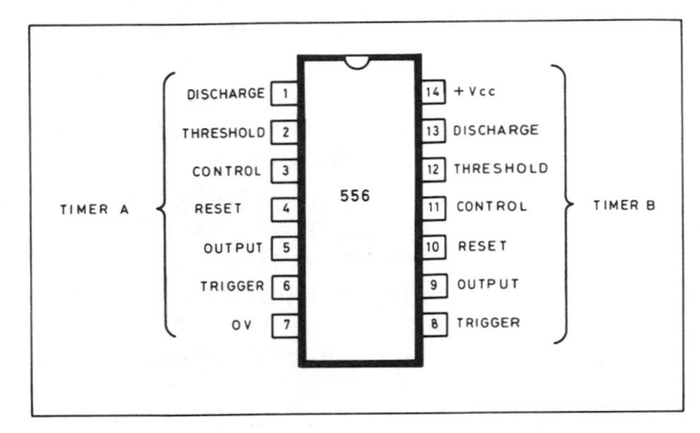

Figure 6.14 Pin connections for the 556 dual timer

circuitry, as shown in Figure 6.12. The 555 comprises two operational amplifiers (used as comparators) together with an RS bistable element. In addition, an inverting output buffer is incorporated so that a considerable current can be sourced to, or sunk from, a load. A single transistor switch, TR1 provides a means of discharging the external timing capacitor. The standard 555 timer is housed in an 8-pin DIL package (see Figure 6.13) and operates from supply rail voltages of between 4.5V and 15V.

555 variants

Dual 555 timer (e.g. NE556A)
Dual 555 in a 14-pin DIL package (see Figure 6.14).

Low power (CMOS) 555 (e.g. ICM7555IPA, TLC555)
Pin-compatible CMOS 555 which operates over a wide supply voltage range. Power consumption is minimal but current drive is very much smaller than that which can be obtained from a standard 555 (up to 200mA).

Low-power (CMOS) dual 555 (e.g. ICM7556IPD, TLC556)
Dual ICM7555IPA (i.e. a low-power pin-compatible CMOS 556).

Hints and tips
★ Timers can be used in either *astable mode* (to generate a continuous pulse train) or in *monostable mode* (to generate a single pulse of accurately defined length). In astable mode, the range of pulse repetition frequencies (prf) that can be reliably obtained extends from less than 0.1Hz (ten seconds per cycle) to over 100kHz. In monostable mode, pulses of duration ranging from 10μs to 10s can be generated. In either case, accuracy will be determined primarily by the external components (resistors and capacitors employed). Outside the above ranges, it is usually necessary to employ alternative techniques (such as monostable logic circuits, e.g. 74121).
★ Low-power timers (e.g. CMOS 555) are not capable of delivering the substantial levels of output current (up to 200mA for a standard 555). CMOS devices will, however, generally source and sink sufficient current to support the connection of up to two TTL loads.
★ Since the 555 timer is capable of switching appreciable currents very rapidly, it is essential to provide adequate supply decoupling. In order to avoid noises and spikes appearing on the supply rail, this decoupling should be provided as close as possible to the 555 device. In most applications, an electrolytic capacitor of 47μF or 100μF (of suitable voltage rating) wired close to pin-8 of a standard 555 (and with an effective ground connection) will eliminate supply borne noise.

Voltage regulators

Integrated circuit regulators are commonly employed in conjunction with stabilised d.c. power supplies. Regulators are invariably of the three-terminal variety and may be either fixed or variable voltage types. Most regulators incorporate internal current limiting and thermal shut-down. Various regulator encapsulations are shown in Figure 6.15.

Voltage regulator data

Fixed voltage regulators

Device	*Output voltage (V)*	*Output current (A)*	*Pinout ref.*
7805	+5	1	a (Fig 6.15)
7905	−5	1	b
7809	+9	1	a
7909	−9	1	b
7812	+12	1	a
7912	−12	1	b
7815	+15	1	a
7915	−15	1	b
7824	+24	1	a
7924	−24	1	b
78L05	+5	0.1	c
79L05	−5	0.1	d
78L12	+12	0.1	c
79L12	−12	0.1	d
78L15	+15	0.1	c
79L15	−15	0.1	d
78S05	+5	2	a
78S12	+12	2	a
78S15	+15	2	a
78S24	+24	2	a
78T05	+5	3	a
78T12	+12	3	a
78T15	+15	5	a
78H05	+5	5	e
78H12	+12	5	e

Variable voltage regulators

Device	*Output voltage*		*Output current (A)*	*Pinout ref.*
	min. (V)	*max. (V)*		
L200	2.9	36	2	
LM317LZ	1.2	37	0.1	
LM317T	1.2	37	1.5	f
LM317K	1.2	37	1.5	g
LM338K	1.2	32	5	g
LM396K	1.2	15	10	h

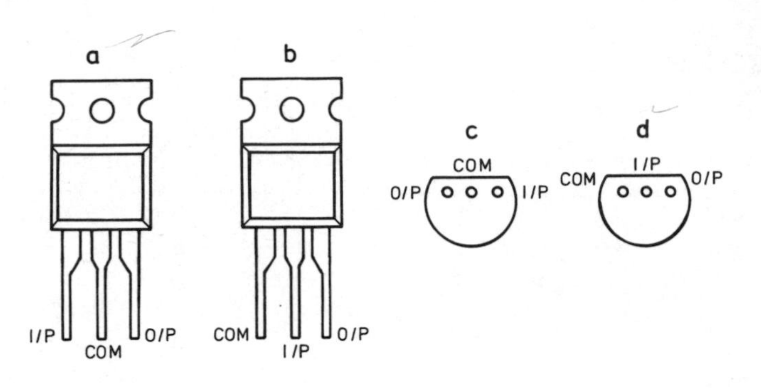

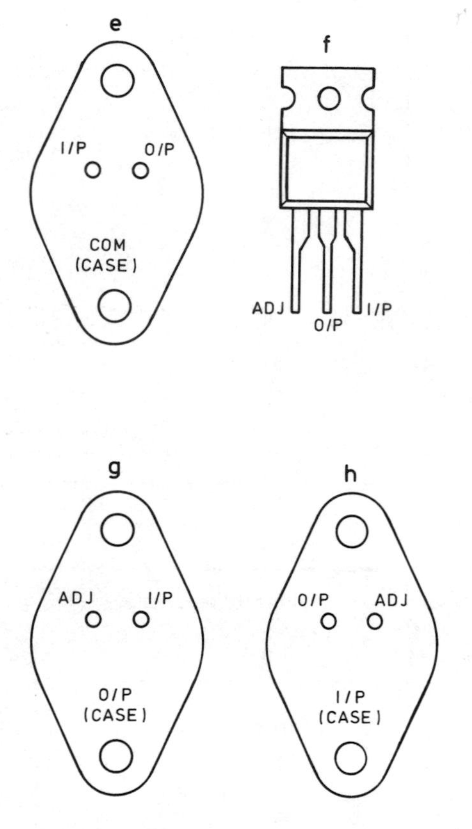

Figure 6.15 Pin connections for common voltage regulators

Hints and tips

★ Voltage regulators have considerable internal gain and thus, to ensure unconditional high-frequency stability, it is recommended that relatively low value (e.g. 100nF) capacitors be fitted (from input to ground and from output to ground) as close as possible to the device.

★ In the vast majority of applications, voltage regulators will operate at significant power levels. It is thus essential to provide adequate heat sinking (following the manufacturer's recommendations). Failure to observe this precaution may result in premature current limiting or output voltage foldback due to thermal shutdown.

★ Care should be taken to ensure that the (unregulated) d.c. input voltage is within the range specified by the manufacturer (e.g. +7V to +25V max. for a 7805). Where the input voltage is very much higher than the output voltage, power dissipation within the regulator may be excessive. Conversely, where the input voltage is only marginally greater than the output voltage, regulation may be very poor and ripple may be present at high current levels. The optimum input voltage (on load) must normally be at least 3V greater than the regulator's output voltage. In order to avoid excessive regulator power dissipation, the input voltage (off load) should normally be no more than about 6V greater than the regulator's output voltage.

7 Circuits

Transistor amplifiers

Three different circuit configurations are employed for transistor amplifiers depending upon which of the three transistor connections is *common* to both input and output (see Figure 7.1 and 7.2 for bipolar and field effect (JFET) transistors respectively. The three basic circuit configurations exhibit the following characteristics (typical values given in brackets):

Bipolar transistors

Parameter	*Mode of operation*		
	Common emitter	*Common collector*	*Common base*
voltage gain	medium/high (40)	unity (1)	high (200)
current gain	high (200)	high (200)	unity (1)
power gain	very high (8000)	high (200)	high (200)
input resistance	medium (2.5kΩ)	high (100kΩ)	low (200Ω)
output resistance	medium/high (20kΩ)	low (100Ω)	high (100kΩ)
phase shift	180°	0°	0°
typical applications	general purpose, AF and RF amplifiers	impedance matching, input and output stages	RF and VHF amplifiers

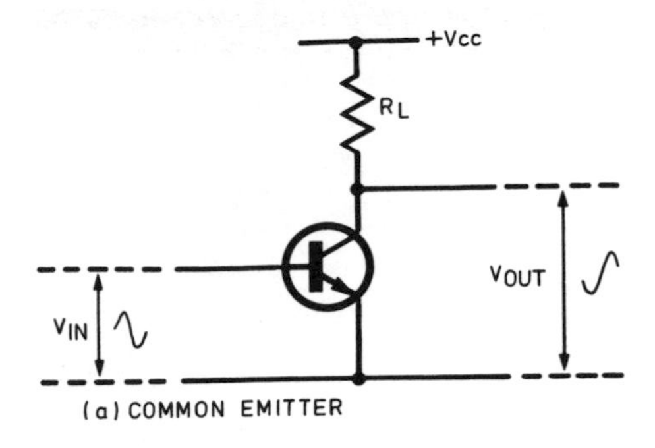

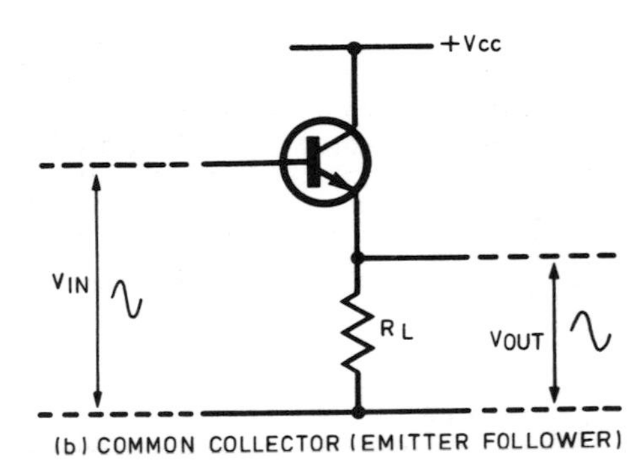

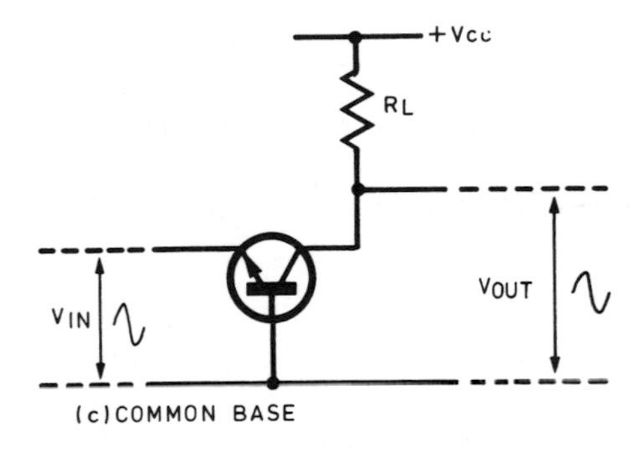

Figure 7.1 Basic circuit configurations for bipolar transistors

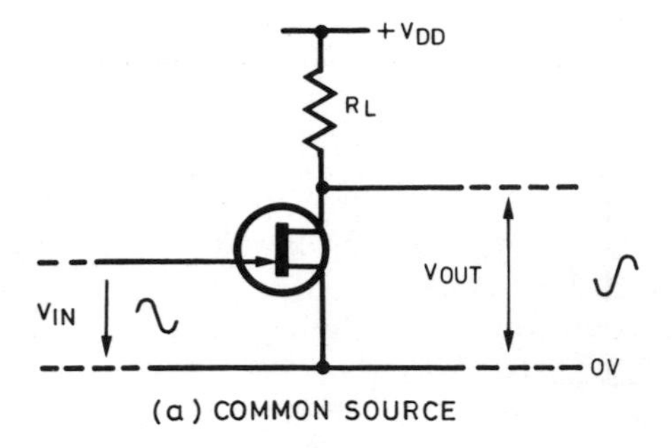

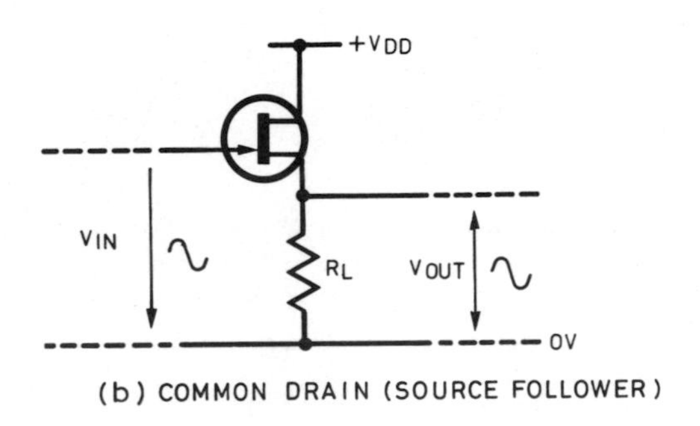

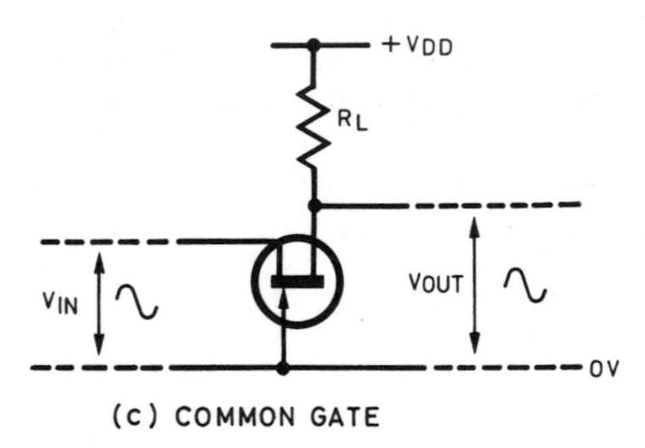

Figure 7.2 Basic circuit configurations for JFET transistors

Field effect transistors

Parameter	*Mode of operation*		
	Common source	*Common drain*	*Common gate*
voltage gain	medium (40)	unity (1)	high (250)
current gain	very high (100000)	very high (200000)	unity (1)
power gain	very high (4000000)	very high (200000)	high (250)
input resistance	very high (1MΩ)	very high (1MΩ)	low (500Ω)
output resistance	medium/high (50kΩ)	low (200Ω)	high (150kΩ)
phase shift	180°	0°	0°
typical applications	general purpose, AF and RF amplifiers	impedance matching, input and output stages	RF and VHF amplifiers

Figure 7.3 shows outline circuits for a bipolar transistor operating in common emitter mode. Figure 7.3(a) shows a basic common emitter amplifier with bias current provided by means of R_B and a collector load provided by R_L. Figure 7.3(b) shows an improved method of bias in which negative feedback (both a.c. and d.c.) is provided by connecting the bias resistor to the collector rather than to the positive supply rail, V_{CC}. This circuit offers greater stability, reduced distortion, and less susceptibility to transistor gain (h_{fe}) variations at the cost of some voltage gain. Figure 7.3(c) shows a common-emitter amplifier with emitter stabilisation. This circuit offers the highest degree of stability coupled with relatively high values of voltage gain (40 to 100 typical) with an input impedance of 2kΩ. The circuit offers a bandwidth of several hundred kHz with typical cut-off frequencies of 10Hz and 250kHz. The following component values are typical:

Transistor:	BC107, BC108, BC109, BC184 (etc)
C_{IN}, C_{OUT}	10μF (rated at 25V)
C_E	100μF (rated at 10V)
R_{B1}	100kΩ
R_{B2}	22kΩ
R_L	3.9kΩ
R_E	1kΩ

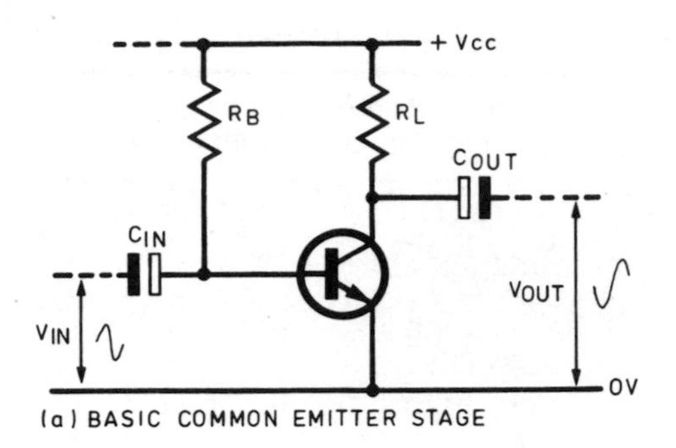

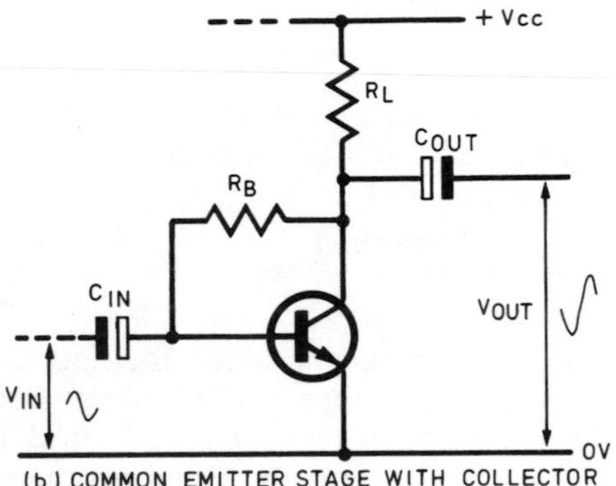

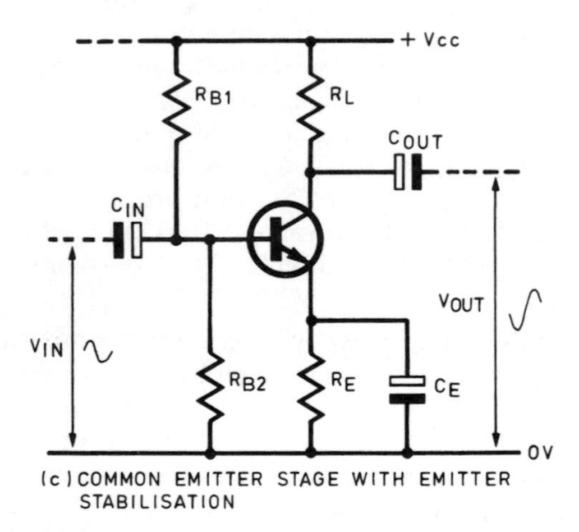

Figure 7.3 Common emitter amplifier circuits

The circuit of Figure 7.3(c) can easily be modified for RF use by replacing the collector load (R_L) with a parallel tuned circuit (or tuned transformer). Maximum voltage gain will be obtained at the frequency of resonance and adequate performance will be achieved over the range 100kHz to over 30MHz (depending upon the values of L and C).

Figure 7.4 shows the outline arrangement of a simple transistor power amplifier. This circuit uses an NPN-PNP complementary

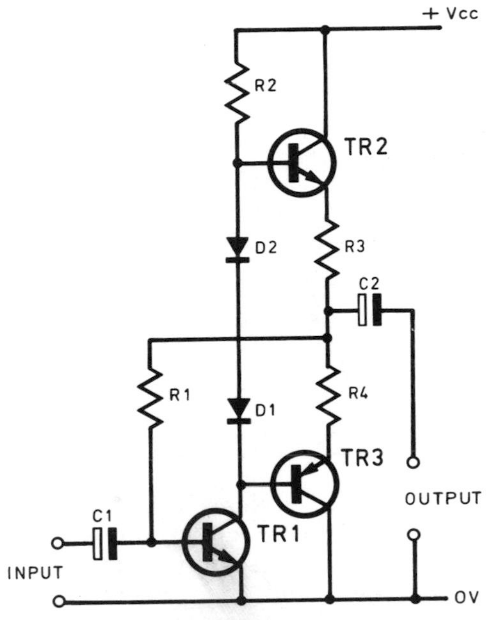

Figure 7.4 Complementary power amplifier stage

pair (TR2 and TR3 respectively) driven by a common emitter amplifier stage (TR1). Negative feedback bias is applied via R1. D1 and D2 provide a constant voltage bias supply for the two output transistors (the voltage drop developed across the two series-connected silicon diodes amounting to 1.2V). R2 is the collector load for TR1, whilst R3 and R4 provide emitter stabilisation for TR2 and TR3. The circuit will provide output powers of up to about 1W into an 8Ω load when operated from a 9V supply. The following component values are typical:

TR1	BC108
TR2	BFY50
TR3	BC461
D_1, D_2	1N4148
C_1	10μF (rated at 15V)
C_2	470μF (rated at 15V)
R_1	330kΩ
R_2	1kΩ
R_3	2.2Ω
R_4	2.2Ω

Alternatively, with an 18V supply and suitably rated transistors, the circuit will provide output powers of up to about 5W into an 8Ω loudspeaker. The following are representative component values:

TR1	BC337
TR2	BD131
TR3	BD132
D_1, D_2	1N4148
C_1	22μF (rated at 25V)
C_2	1000μF (rated at 25V)
R_1	100kΩ
R_2	680Ω
R_3	0.22Ω
R_4	0.22Ω

In order to balance the quiescent voltage developed across the two output transistors, TR2 and TR3, R_1 can be replaced by a fixed resistor of 47kΩ connected in series with a preset resistor of 100kΩ. This component should be adjusted until the voltage at the junction of R3 and R4 is exactly half that of the supply (i.e. $V_{CC}/2$). TR2 and TR3 should be fitted with a heat sink of 10°C/W, or better.

The maximum undistorted output power which can be delivered by a complementary output stage of the type shown in Figure 7.4 is given by the formula:

$$P_{OUT}\ \text{max} = \frac{V_{CC}^2}{8\ R_L}\ \text{W}$$

where V_{CC} is the supply voltage and R_L is the impedance of the load (loudspeaker).

Example 7.1
A complementary output stage operates from a 25V supply rail into a load having an impedance of 8Ω. Determine the maximum possible undistorted output power.

Here V_{CC} = 25V and R_L = 8Ω hence:

$$P_{OUT}\ max = \frac{25^2}{8 \times 8} = \frac{625}{64} = 9.76W$$

Hints and tips
★ Transistor current gain (h_{fe}) is liable to considerable variation and it is wise to design circuits that are tolerant of a wide variation in current gain. In the case of linear circuits (e.g. amplifiers) negative feedback should be used to determine actual stage gains. Remember, also, that current gain falls at high values of collector current.
★ Low noise transistors (e.g. BC109) should be used in circuits where signals have a relatively low amplitude (100mV, or less). Such devices should normally be fitted in the first and subsequent stages of an audio pre-amplifier.
★ Complementary output stages can be improved by using power MOSFETs as the output transistors. Such devices are less prone to thermal instability than their bipolar counterparts.

Transistor oscillators

Simple forms of transistor oscillator (each based on a conventional common emitter transistor amplifier stage) are shown in Figure 7.5. The circuit of Figure 7.5(a) shows a Twin-T network oscillator arrangement. The output of this arrangement comprises a sine wave at a frequency given by:

$$f_{OUT} = \frac{1}{2\pi C R}$$

where $R = R_1 = R_2 = 2R_3$ and $C = C_1 = C_2 = C_3/2$

Typical values for this arrangement are as follows:

Transistor	BC109, BC182, BC184 (etc)
C_1, C_2	10nF
C_3	22nF
R_1, R_2	10kΩ
R_3	4.7kΩ

R_L 3.9kΩ
C_4 1μF

With a supply voltage ($+V_{CC}$) of 9V, the output voltage is a sine wave of approximately 2V pk-pk at 1.6kHz. The frequency may be adjusted by about ±10% by placing a preset resistor of appropriate value in series with R_3.

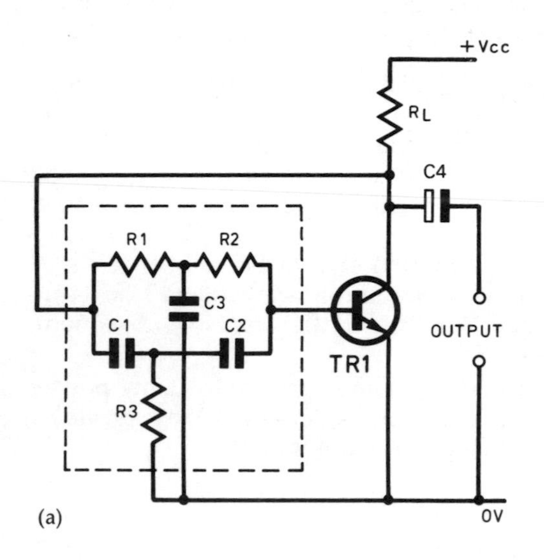

(a)

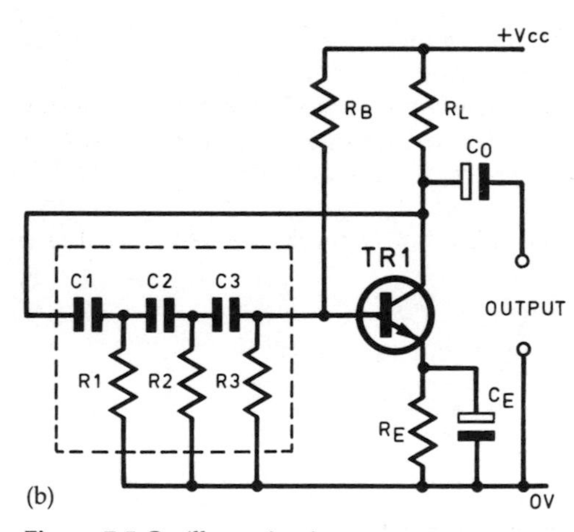

(b)

Figure 7.5 Oscillator circuits

Figure 7.5(b) shows a ladder network oscillator. This arrangement is based on a phase shift of 60° occurring in each of the C-R section within the ladder network. The output frequency is given by:

$$f_{OUT} = \frac{1}{2\,\pi\,\sqrt{6}\,C\,R}$$

where $R = R_1 = R_2 = R_4$ (R_4 is the effective resistance of the parallel combination of R_3, R_B, and the input resistance of TR1), and $C = C_1 = C_2 = C_3$.

Typical values for this arrangement are as follows:

Transistor	BC109, BC182, BC184 (etc)
C_1, C_2, C_3	10nF
C_E	10µF
C_O	1µF
R_1, R_2	2.2kΩ
R_3	4.7kΩ
R_B	39kΩ
R_L	3.3kΩ
R_E	1kΩ

With a supply voltage ($+V_{CC}$) of 9V, the output voltage is a sine wave of approximately 1V pk-pk at 2.8kHz.

Example 7.2

A sine wave signal at 1kHz is required. Devise a suitable circuit arrangement.

Here we shall adopt a twin-T oscillator arrangement as this circuit is generally less temperamental and is easier to trim than an arrangement based upon a three section phase-shift ladder network.

Assuming a value for R of 4.7kΩ, the requisite value for C can be calculated from:

$$C = \frac{1}{2\,\pi\,R\,f_{OUT}}$$

$$\text{hence } C = \frac{1}{6.28 \times 4.7\text{k}\Omega \times 1\text{kHz}}\ \text{F}$$

$$C = \frac{1}{6.28 \times 4.7 \times 10^3 \times 1 \times 10^3}\ \text{F}$$

$$\text{thus} \quad C = \frac{1}{29.5 \times 10^6} = 0.0339 \times 10^{-6}\ \text{F}$$

$$\text{or} \quad C = 33.9\ \text{nF}$$

The nearest preferred value is 33nF and hence the circuit of Figure 7.5(a) should be used with the following values:

R1 = 4.7kΩ
R2 = 4.7kΩ
R3 = 2.2kΩ
C1 = 33nF
C2 = 33nF
C3 = 68nF

If precise frequency trimming is required, R3 should be replaced by a fixed resistor of 1.8kΩ connected in series with a preset resistor of 1kΩ.

Timer circuits

Timers can be used in monostable or astable mode as generators of accurately defined pulses or of continuous pulse trains. The following terminology is applied to the signals generated by such devices:

Pulse repetition frequency (prf) of a pulse waveform is the number of pulses which occur in a given interval of time (invariably one second).

The *period* of a pulse waveform is the time taken for one complete cycle (on time plus off time) of the pulse. The pulse period is also equal to the reciprocal of the prf, hence:

$$t_P = t_{ON} + t_{OFF} = 1/f_P$$

where t_P is the period of the pulse, t_{ON} is the on (or high) time, t_{OFF} is the off (or low) time, and f_P is the prf of the pulse train.

The *mark to space ratio* of a pulse waveform is the ratio of on (or high) time to off (or low) time. Thus:

$$\text{mark to space ratio} = t_{ON}/t_{OFF}$$

The *duty cycle* of a pulse waveform is the ratio of on (or high) time to on (or high) plus off (or low) times. Hence:

$$\text{duty cycle} = \frac{t_{ON}}{t_{ON} + t_{OFF}}$$

Duty cycle is often expressed as a percentage, i.e.:

$$\text{duty cycle} = \frac{t_{ON}}{t_{ON} + t_{OFF}} \times 100\%$$

The *pulse width* of a rectangular waveform is the time interval (measured at the 50% amplitude points) for which the pulse is on or high. In the case of a perfect rectangular pulse, the pulse width is simply equal to the on time, t_{ON}.

The pulse rise time is equal to the time interval between the 10% and 90% amplitude points of the pulse. The rise time of a perfect rectangular pulse would be zero. Similarly, pulse fall time is the time interval between the 90% and 10% amplitude points of the pulse. Again, the fall time of perfect rectangular wave would be zero.

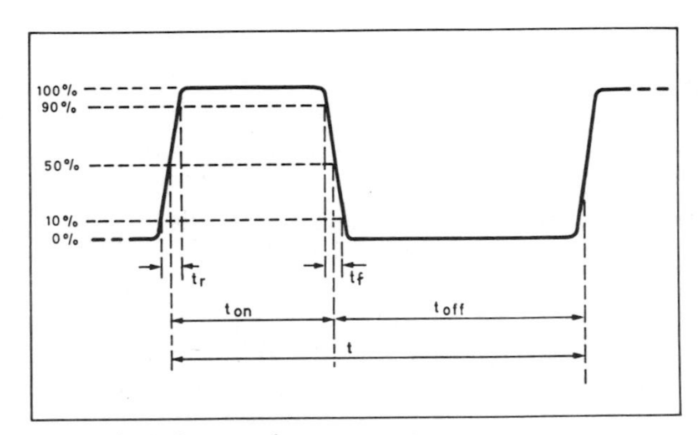

Figure 7.6 Pulse waveform parameters

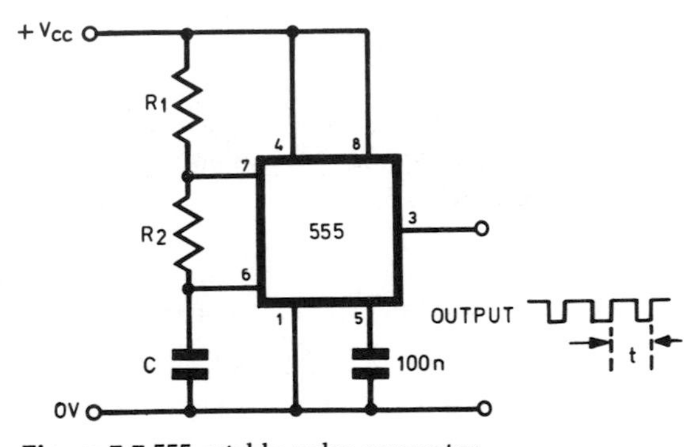

Figure 7.7 555 astable pulse generator

Figure 7.6 illustrates a typical pulse waveform on which these various parameters have been indicated.

Figure 7.7 shows how the standard 555 can be used as an *astable pulse generator*. The output waveform has a period determined by C, R1 and R2 and has an amplitude which is approximately equal to the supply voltage. The output waveform has the following characteristics:

time for which output is high: $t_{ON} = 0.693\,(R1 + R2)\,C$

time for which output is low: $t_{OFF} = 0.693\,R2\,C$

period of output: $t_F = t_{ON} + t_{OFF} = 0.693\,(R1 + 2R2)\,C$

prf of output: $f_P = \dfrac{1.44}{(R1 + 2R2)\,C}$

mark to space ratio of output: $t_{ON}/t_{OFF} = \dfrac{R1 + R2}{R2}$

duty cycle of output: $\dfrac{t_{ON}}{t_{ON} + t_{OFF}} = \dfrac{R1 + R2}{R1 + 2R2} \times 100\%$

(NB: The nomograph shown in Figure 7.8 can be used to determine the prf, f_P, in astable mode).

Figure 7.9 shows the 555 connected as a *monostable pulse generator*. The monostable timing period is initiated by a falling edge (i.e. a high to low transition) applied to the trigger input. When such an edge is received and the trigger input voltage falls to below 1/3 of the supply voltage, the 555 output (pin-3) goes high and the monostable pulse is generated for a period determined by the time constant $C \times R$. The amplitude of the pulse will be approximately equal to the supply voltage. The output pulse has the following characteristics:

period for which the output is high: $t_{ON} = 1.1\,R\,C$

recommended trigger pulse width: $t_{TR} < 0.25\,t_{ON}$

(NB: The nomograph shown in Figure 7.10 can also be used to determine the monostable pulse width, t_{ON}).

Example 7.3

A square wave having an approximate frequency of 400Hz and mark to space ratio of 2:1 (duty cycle 67%) is required. If the waveform is to have an amplitude of 5V, devise a suitable circuit arrangement.

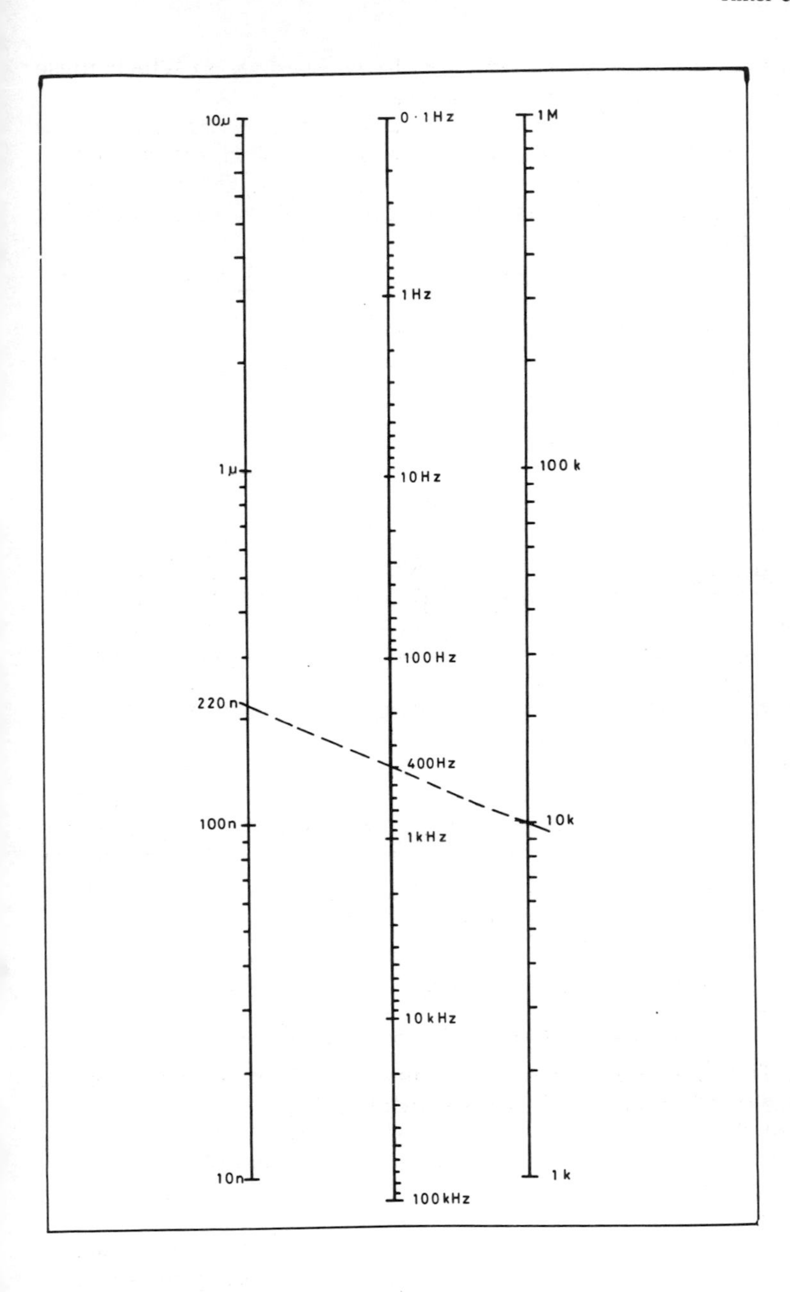

Figure 7.8 Nomograph for the 555 astable pulse generator

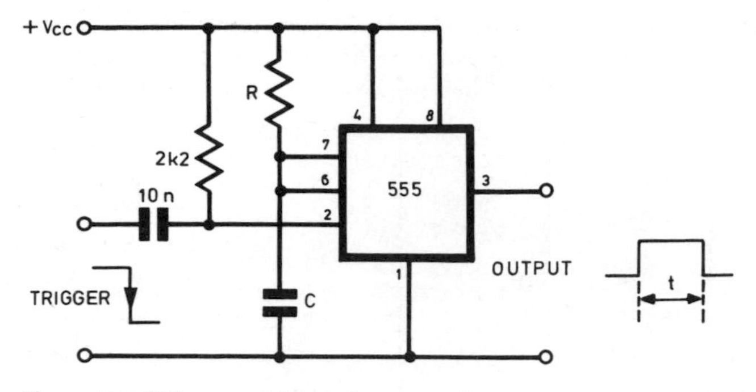

Figure 7.9 555 monostable pulse generator

Here we require a 555 connected in astable mode, as shown in Figure 7.7. Since the duty cycle is 2:1, R1 must be equal to R2. Assuming that we choose a value for R1 of 10kΩ, we can determine the value of C using the nomograph of Figure 7.8. At 400Hz, the nearest preferred value of capacitance is 220nF (see Figure 7.8). Hence the astable circuit of Figure 7.7 should be employed with:

R1 = 10kΩ
R2 = 10kΩ
C = 220nF
V_{CC} = +5V

Example 7.4
A pulse is to be generated in response to a falling edge of amplitude 5V. If the pulse is to have a duration of approximately 5ms, devise a suitable circuit arrangement.

Here we require a 555 connected in monostable mode, as shown in Figure 7.9. Assuming that we choose a value for C of 100n, we can determine the value of R using the nomograph of Figure 7.10. In order to produce a 5ms pulse, the nearest preferred value of resistance is 47kΩ. Hence the monostable circuit of Figure 7.9 should be employed with:

R = 47kΩ
C = 100nF
V_{CC} = +5V

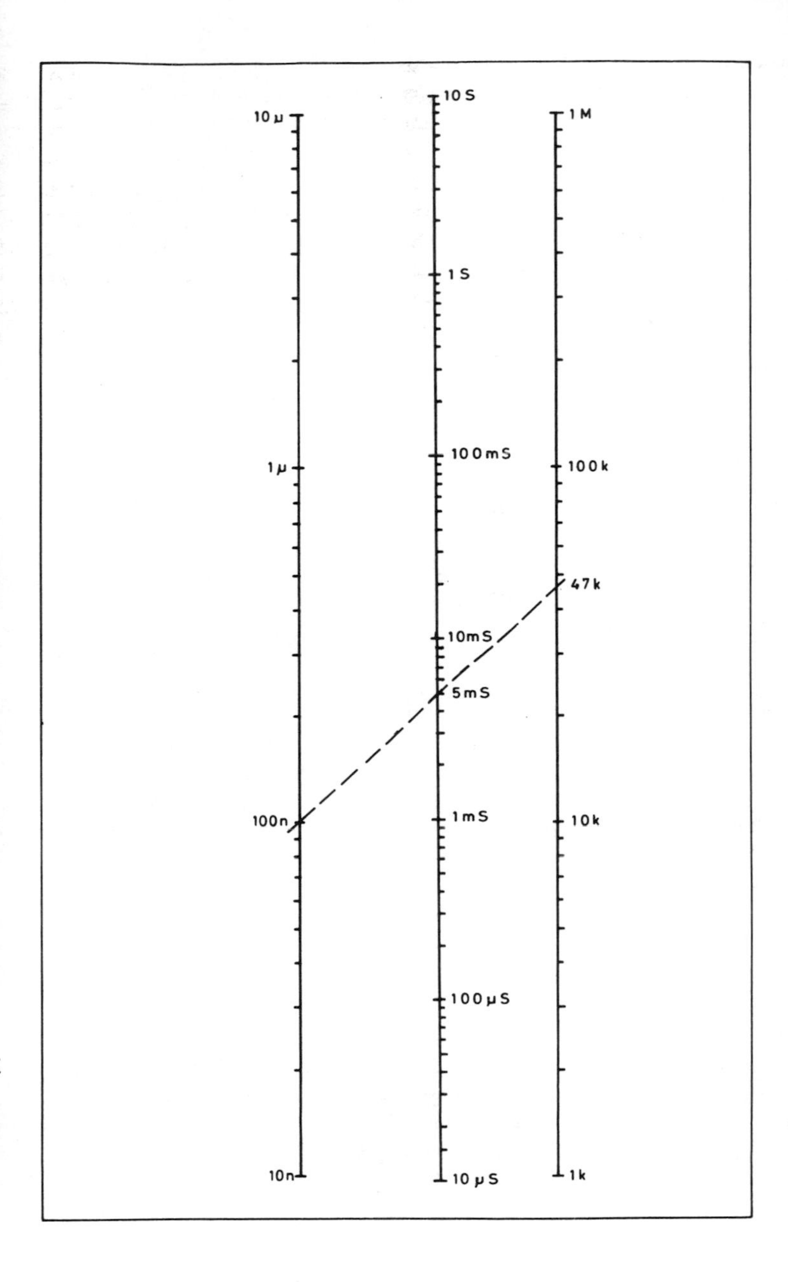

Figure 7.10 Nomograph for the 555 monostable pulse generator

Example 7.5

A square wave having an approximate frequency of 1kHz and equal mark to space ratio is required. If the waveform is to have an amplitude of 12V, devise a suitable circuit arrangement.

A 555 connected in astable mode will be required (Figure 7.7). Since the duty cycle is to be 1:1, we need to make $R1 \ll R2$. Assuming that we use the minimum recommended value for R1 (1kΩ) a suitable value for R2 would be 100 times this value, or 100kΩ. The value of capacitance can be calculated from:

$$\text{prf of output:} C = \frac{1.44}{(R1 + 2R2)\, f_P}$$

hence

$$C = \frac{1.44}{(1k\Omega + 200k\Omega) \times 1kHz} \text{ F}$$

$$C = \frac{1.44}{201 \times 10^3 \times 1 \times 10^3} \text{ F}$$

$$C = \frac{1.44}{201} \times 10^{-6} \text{ F}$$

thus

$$C = 7nF$$

In practice a 6.8nF would probably be adequate. If the frequency requires accurate trimming, R2 could be replaced by a fixed resistor of 68kΩ connected in series with a preset resistor of 47kΩ. Finally, in order to obtain an output of approximately 12V, a supply voltage (V_{CC}) of 12V should be used.

Heatsinks

Heatsinks should be fitted to any transistor or integrated circuits which is likely to dissipate an appreciable level of power. Heatsinks are available in various styles (to permit mounting of different semiconductor case styles) and in various ratings (to cater for different levels of power dissipation).

The *thermal resistance* (expressed in °C/W) is a measure of its efficiency as a dissipator of heat. When a system has reached thermal equilibrium, the power dissipated by the semiconductor device is equal to the ratio of temperature difference (between junction and ambient) to the sum of the thermal resistance present. Hence,

$$\text{power dissipated, } P_{DISS} = \frac{T_J - T_A}{\theta_{JC} + \theta_{CS} + \theta_{SA}}$$

where T_J is the maximum junction temperature (in °C) specified by the manufacturer (*before* derating is applied), T_A is the ambient temperature (in °C), θ_{JC}, θ_{CS}, and θ_{SA} are the thermal resistance of junction to case (specified by the semiconductor manufacturer), case to heatsink, and heatsink to air respectively (in °C/W).

The formula may be arranged as follows:

$$\theta_{SA} = \frac{T_J - T_A}{P_D} - (\theta_{JC} + \theta_{CS}) \text{ °C/W}$$

Example 7.6

A transistor has the following ratings:

Maximum junction temperature: 95°C
Thermal resistance (junction to case): 2°C/W

The transistor is to dissipate 10W in a circuit where the ambient temperature is not expected to exceed 40°C. Determine the thermal resistance of the heatsink which must be fitted.

Assuming that the thermal resistance of the mounting hardware (θ_{CS}) is 1°C/W (a typical figure) the required thermal resistance will be given by:

$$\theta_{SA} = \frac{95 - 40}{10} - (2 + 1) \text{ °C/W}$$

thus $\theta_{SA} = 5.5 - 3 = 2.5$ °C/W

Hence a heatsink rated at 2.5 °C/W (or better) should be fitted.

Hints and tips

★ The thermal resistance from case to sink (θ_{CS}) can be signifi cantly reduced by the application of a thin film of heatsink compound. This material combines excellent thermal conductivity with a very high dielectric resistance. Contamination risk can be reduced by using compounds which are based on synthetic fluids rather than silicone.

★ Silicone impregnated rubber insulating washers can be fitted to power semiconductors. The typical thermal resistance of these washers ranges from 0.4°C/W for the TO3 variety to 3°C/W for TO126 types.

★ Consideration should be given to the location of a heatsink so that efficient radiation and convection of heat can take place. Heatsinks should be matt black finished and placed where unrestricted convection air currents can flow.

Power supply circuits

Most electronic circuits require a source of well regulated d.c. at voltages of between 5V and 25V. Some circuits (e.g. those based on operational amplifiers) also require symmetrical supply rails of typically ±9V, ±12V, or ±15V.

An unregulated low-voltage d.c. supply can be obtained using the transformer-rectifier arrangements shown in Figure 7.11.

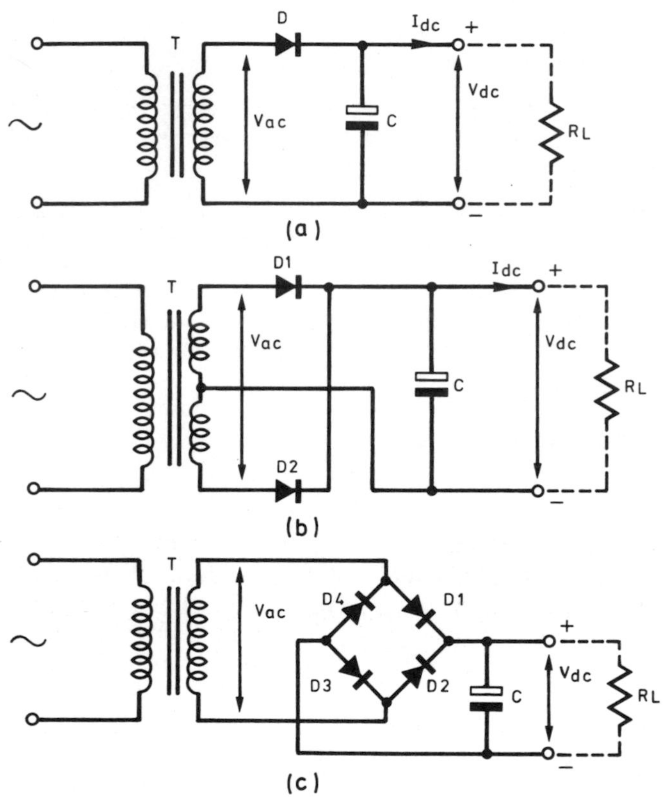

Figure 7.11 Transformer-rectifier arrangements

Figure 7.11(a) employs *half-wave* rectification and is unsuitable for load-currents much in excess of 100mA. Figure 7.11(b) is a split-secondary transformer in conjunction with two rectifier diodes connected in a *bi-phase* arrangement. Figure 7.11(c) shows the most common *full-wave* circuit configuration which is based on a four-diode bridge. The following formulae apply:

	Half-wave	*Full-wave*	
		Bi-phase	*Bridge*
Circuit	Fig. 7.11(a)	Fig. 7.11(b)	Fig. 7.11(c)
Output voltage (V_{dc})	$1.41 \times V_{ac}$	$0.71 \times V_{ac}$	$1.41 \times V_{ac}$
Output current (I_{dc})	$0.28 \times I_{ac}$	I_{ac}	$0.62 \times I_{ac}$
Input voltage (V_{ac})	$0.71 \times V_{dc}$	$1.41 \times V_{dc}$	$0.71 \times V_{dc}$
Input current (I_{ac})	$3.57 \times I_{dc}$	I_{dc}	$1.61 \times I_{dc}$
Recommended min. reservoir capacitance (μF if I_{dc} in A)	$4700 \times I_{dc}$	$2200 \times I_{dc}$	$2200 \times I_{dc}$

In each case the *reservoir capacitor* should be rated at around $2 \times V_{dc}$ in order to provide an adequate safety margin.

Most power supplies will involve the use of an integrated circuit regulator which will provide *ripple rejection* (smoothing) and *voltage regulation*. Most regulators will require an input voltage of at least 2V greater than their nominal output voltage at full load. In order to avoid excessive power dissipation within the regulator, the regulator input voltage should not be too great. A practical overhead voltage (V_{OH}) for a regulator is thus in the region 3V to 6V. Hence:

$$3 \le V_{OH} \le 6$$

and since $V_{OH} = V_{dc} - V_{REG}$

thus $V_{dc} + 3 \le V_{REG} \le V_{dc} + 6$

where V_{dc} is the d.c. output voltage from the rectifier arrangement (measured across the smoothing capacitor) and V_{REG} is the regulated output voltage (i.e. the nominal voltage provided by the regulator). Both V_{dc} and V_{REG} are measured in volts under full-load conditions. The full-load power dissipation in the regulator will be given by:

$$P_{REG} = I_{dc} \times V_{OH} = I_{dc} \times (V_{dc} - V_{REG})$$

Example 7.7
A regulated power supply capable of supplying 12V at 1A is required. Devise a suitable circuit arrangement.

Here V_{REG} = 12V and I_{dc} = 1A. Now:

$$12 + 3 \leq V_{dc} \leq 12 + 6$$

A suitable value for the regulator input voltage will thus be in the range 15V (min.) to 18V (max.). Assuming that a full-wave bridge rectifier arrangement is used, a suitable value for the transformer secondary voltage can now be determined. Let us assume that we aim for an unregulated d.c. voltage of 17V (i.e. 1V less than the maximum recommended) then:

$$V_{ac} = 0.71 \times V_{dc} = 0.71 \times 17V = 12.07V.$$

Hence a transformer having a 12V secondary winding will be required. The transformer secondary should be rated at:

$$I_{ac} = 1.61 \times I_{dc} = 1.61A$$

In practice, a 2A secondary will be more than adequate. The power dissipated by the regulator will be:

$$P_{REG} = 1A \times (17V - 12V) = 5W$$

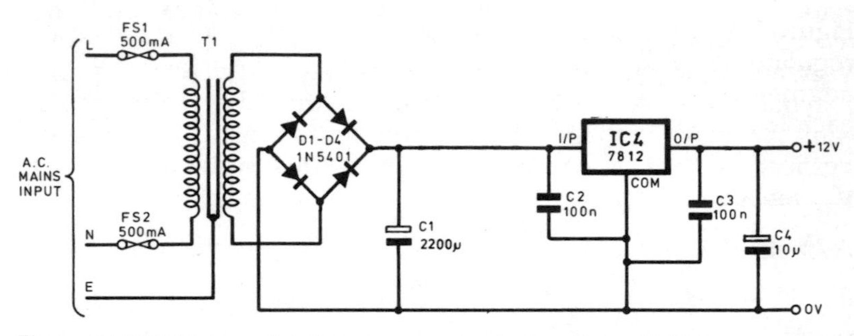

Figure 7.12 12V 1A regulated d.c. power supply (see Example 7.7)

A heatsink rated at 4°C/W, or better will be required. The circuit of this arrangement is shown in Figure 7.12.

Example 7.8
Separate power supplies of +12V and −12V at 5A are required. Devise a suitable circuit arrangement.

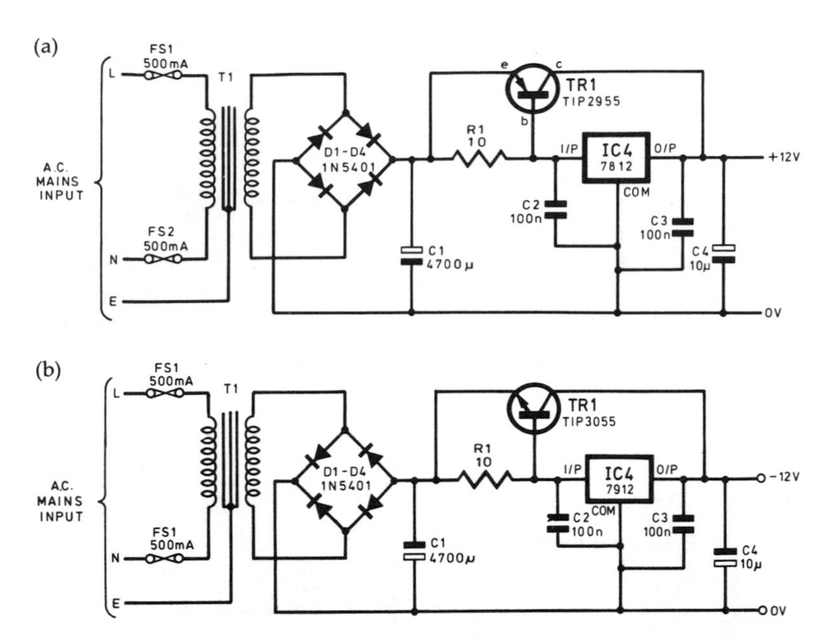

Figure 7.13 Use of a current boosting transistor (see Example 7.8)

Figure 7.13(a) and (b) respectively show circuits based on standard regulators (+12V and −12V). The voltage regulators are augmented (current boosted) by low-cost power transistors. In each circuit arrangement the dissipation in the current boost transistor will be given by:

$P_{REG} = 5A \times (17V - 12V) = 25W$

A heatsink rated at 1.1°C/W, or better will be required.

D.C. to d.c. converters

The 555 timer can be used in simple d.c. to d.c. converter arrangements, as shown in Figure 7.14. The circuit of Figure 7.14(a) produces a positive output voltage which is approximately double that of the input, V_{IN}. The circuit of Figure 7.14(b) produces a negative output voltage which is approximately the same as the input voltage, V_{IN}, but of opposite sign. Note that the maximum current that should be drawn from either of these simple voltage converters is limited to about 10mA.

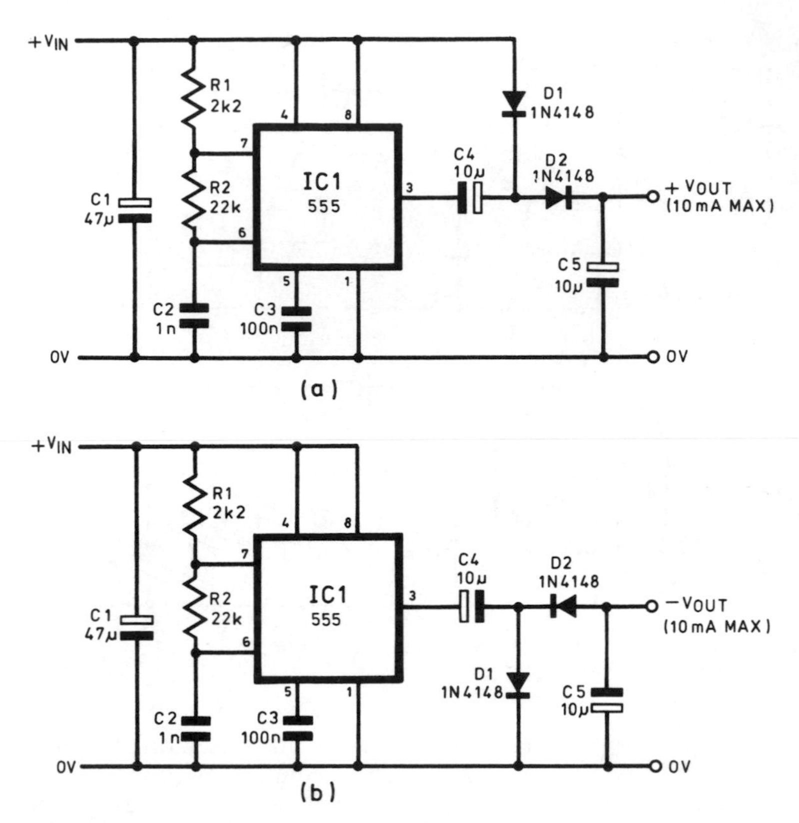

Figure 7.14 D.C. to d.c. converters based on a 555

Operational amplifier circuits

Operational amplifiers have a wide variety of applications in electronic circuits. Figure 7.15 shows basic amplifier configurations using an operational amplifier. The circuit of Figure 7.15(a) is an *inverting amplifier* which provides a phase shift of 180° between the input and output signals. The circuit of Figure 7.15(b) shows a *non-inverting amplifier* which provides a phase shift of 0° (i.e. input and output signals are in-phase). The voltage gain (A_v) of the circuit in Figure 7.15(a) is given by:

$$A_v = \frac{V_{OUT}}{V_{IN}} = \frac{-R_F}{R_{IN}}$$ (the minus sign indicates inversion not attenuation)

(a)

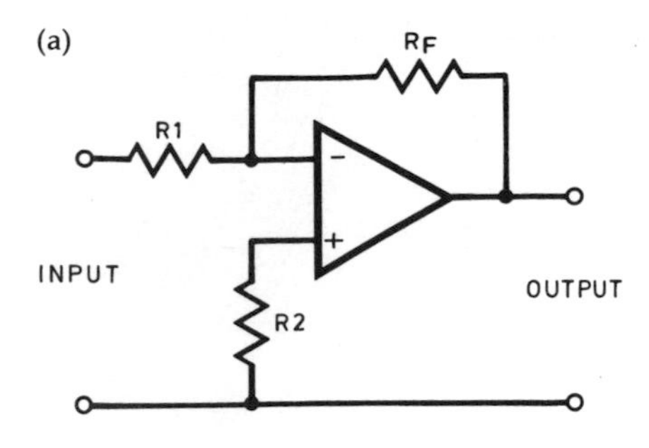

(b)

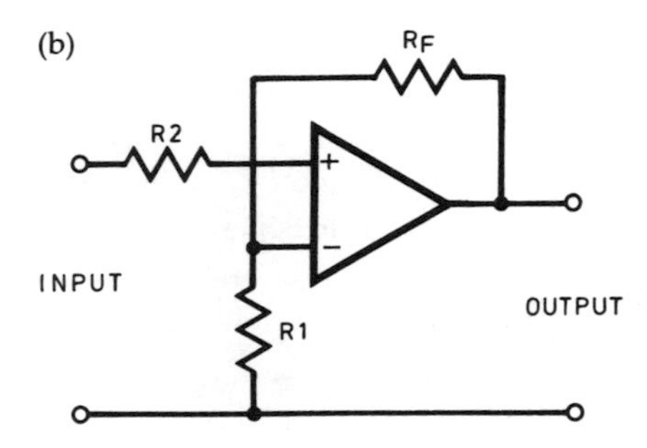

Figure 7.15 Basic amplifier stages

whilst that for Figure 7.15(b) is given by:

$$A_v = \frac{V_{OUT}}{V_{IN}} = \frac{R_F}{R_{IN}}$$

For optimum performance, R_2 is equal to the parallel combination of R_1 and R_F in order to retain symmetry. Hence (for both circuits):

$$R_2 = \frac{R_1 \times R_F}{R_1 + R_F}$$

Figure 7.16 shows a *summing amplifier* in which three input voltages are summed. Circuits of this type are often used for audio frequency mixers. The output voltage of Figure 7.16 is given by:

$$V_{OUT} = -R_F \times \left\{ \frac{V_{IN1}}{R_1} + \frac{V_{IN2}}{R_2} + \frac{V_{IN3}}{R_3} \right\}$$

Figure 7.17 shows a unity gain buffer stage. This stage offers a voltage gain of 1 coupled with a very high input resistance and

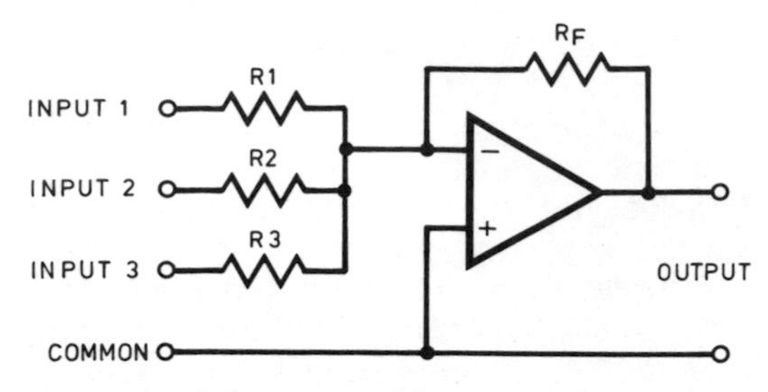

Figure 7.16 Summing amplifier

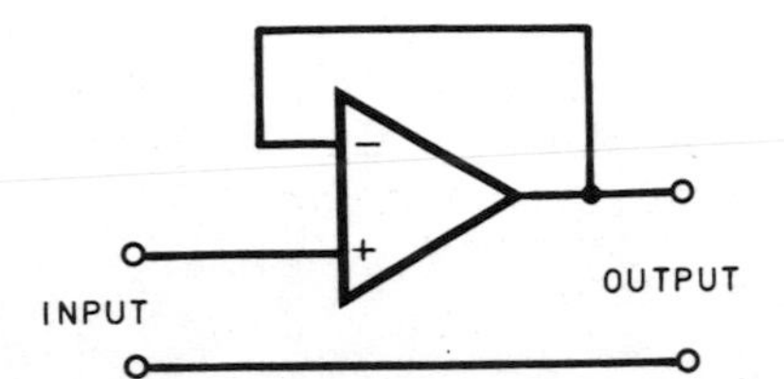

Figure 7.17 Unity gain buffer

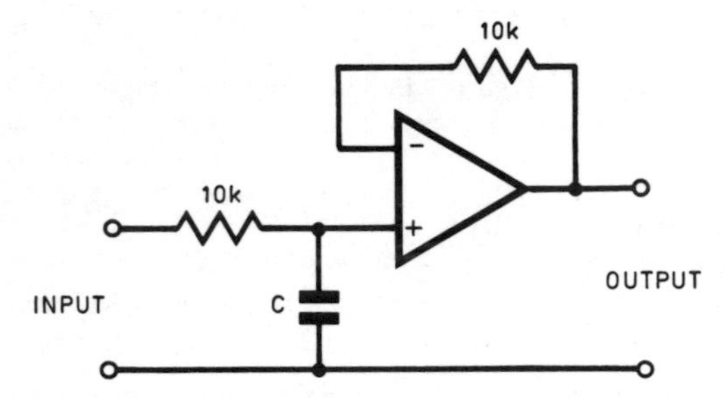

Figure 7.18 Sample and hold based on an operational amplifier

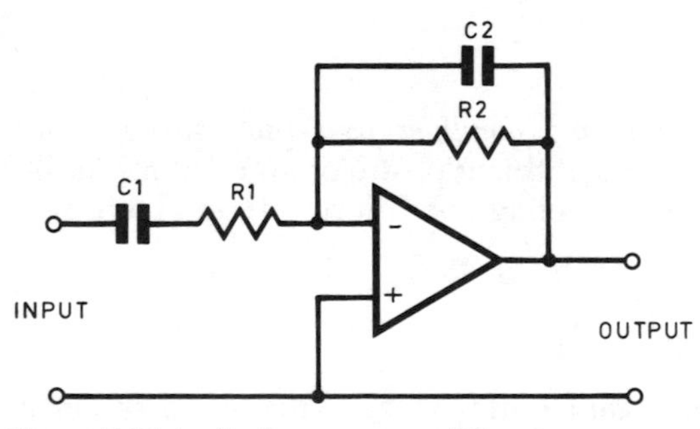

Figure 7.19 Audio frequency amplifier stage

low output resistance. Stages of this type are often used for impedance matching.

Figure 7.18 acts as a simple sample and hold arrangement. If an input voltage is applied, it will be transferred to the capacitor, C. If the input voltage is then removed (i.e. input disconnected), the voltage developed across the capacitor will be maintained for a very long period as its only discharge path is through the very high input impedance of the operational amplifier. The output voltage will be the same as that which appears across the capacitor however the output can be safely connected to a relatively low resistance circuit without the danger of losing the charge stored within the capacitor.

Figure 7.19 shows an audio frequency amplifier which has cut-off frequencies and bandwidth determined by the component values employed. The following equations relate to this circuit:

mid-band voltage gain: $A_{V(MB)} = R_2/R_1$

mid-band input impedance: $Z_{IN(MB)} = R_1$

lower cut-off frequency: $f_{C(LOW)} = \dfrac{1}{2\,\pi\,C_1\,R_1}$ Hz

upper cut-off frequency: $f_{C(HIGH)} = \dfrac{1}{2\,\pi\,C_2\,R_2}$ Hz

bandwidth: $f_W = f_{C(HIGH)} - f_{C(LOW)}$ Hz

Figure 7.20 shows the frequency response of this circuit.

Figure 7.21(a) shows an *active low-pass filter* based on an operational amplifier. This circuit passes signals below the cut-off

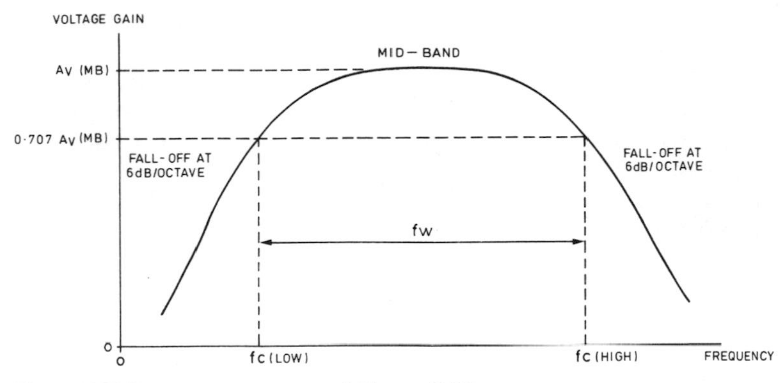

Figure 7.20 Frequency response of Figure 7.19

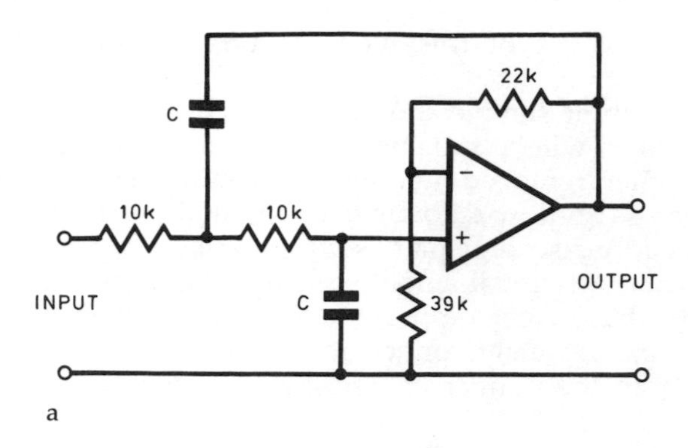

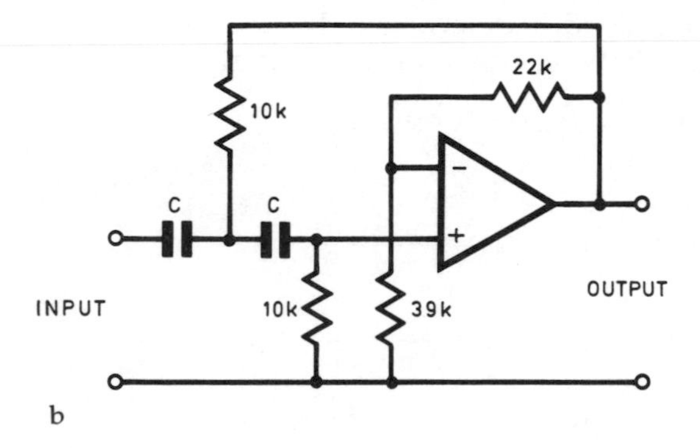

Figure 7.21 Filters based on operational amplifiers

frequency with minimal attenuation. Above the cut-off frequency, response falls off at a rate of 6dB/octave.

Figure 7.21(b) shows an *active high-pass filter* based on an operational amplifier. This circuit passes signals above the cut-off frequency with minimal attenuation. Below the cut-off frequency, response falls off at a rate of 6dB/octave.

The cut-off frequency of the circuits in Figure 7.21 can be calculated from:

$$f_C = \frac{15915}{C} \text{ Hz}$$

where C is expressed in nF.

Example 7.9
An inverting amplifier with a gain of 1 and an input resistance of 10kΩ is required. Devise a suitable circuit arrangement.

Since an inverting amplifier (phase shift = 180°) is specified, the required circuit must follow that shown in Figure 7.15(a). The input resistance of the circuit (R_2) must be 10kΩ hence the feedback resistance (R_F) must also be 10kΩ in order to provide a voltage gain of 1. The value of R_2 should be the equivalent of two 10kΩ resistors

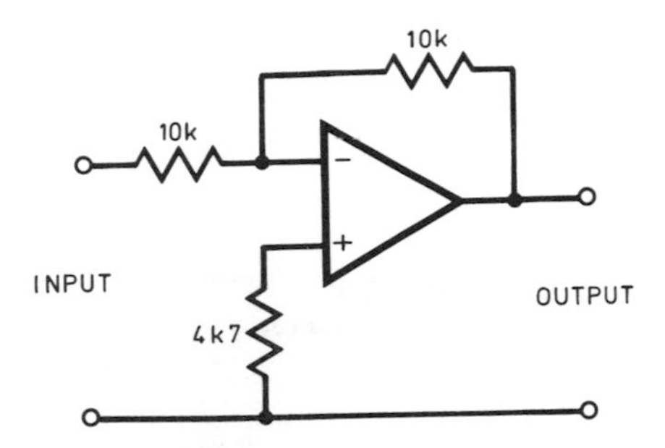

Figure 7.22 Unity gain inverting amplifier stage (see Example 7.9)

in parallel (i.e. 5kΩ), however a preferred value 4.7kΩ resistor will be perfectly adequate. The required circuit arrangement is shown in Figure 7.22.

Example 7.10
An audio amplifier stage is required to have the following specifications:
Mid-band voltage gain = 10
Mid-band input impedance = 10kΩ
Lower cut-off frequency = 50Hz
Upper cut-off frequency = 5kHz
Devise a suitable circuit arrangement.

Since the mid-band input impedance is to be 10kΩ, R_1 must be 10kΩ. In order to provide a mid-band voltage gain of 10, R_2 must be ten times 10kΩ (or 100kΩ). The values of C_1 and C_2 may now be calculated:

$$C_1 = \frac{1}{2\,\pi\,R_1\,f_{C(LOW)}}$$

$$\text{hence } C_1 = \frac{1}{6.28 \times 10 \times 10^3 \times 50} = 318 \times 10^{-9}\text{ F}$$

thus $C_1 = 318nF$ (nearest preferred value 330nF)

Also, $C_2 = \dfrac{1}{2\,\pi\,R_2\,f_{C(HIGH)}}$

hence $C_2 = \dfrac{1}{6.28 \times 100 \times 10^3 \times 5 \times 10^3} = 318 \times 10^{-12}$ F

thus $C_2 = 318pF$ (nearest preferred value 330pF)

The circuit of Figure 7.19 should be used with:

$R_1 = 10k\Omega$, $R_2 = 100k\Omega$, $C_1 = 330nF$, and $C_2 = 330pF$

Example 7.11

A variable gain amplifier stage is required. The adjustment range is to vary from 1 to 10 (approx.). Devise a suitable arrangement.

Stage gain may be made adjustable by including a variable resistor in the feedback path. Assuming that the input resistance of the stage is to be 10kΩ, the feedback resistance will need to be adjustable from 10kΩ to 100kΩ (or more). This can be easily achieved by wiring a fixed resistor of 10kΩ in series with a variable resistor of 100kΩ, as shown in Figure 7.23. Figure 7.24 shows an alternative (non-inverting) arrangement. Both circuits provide a voltage gain which is adjustable from 1 to 11.

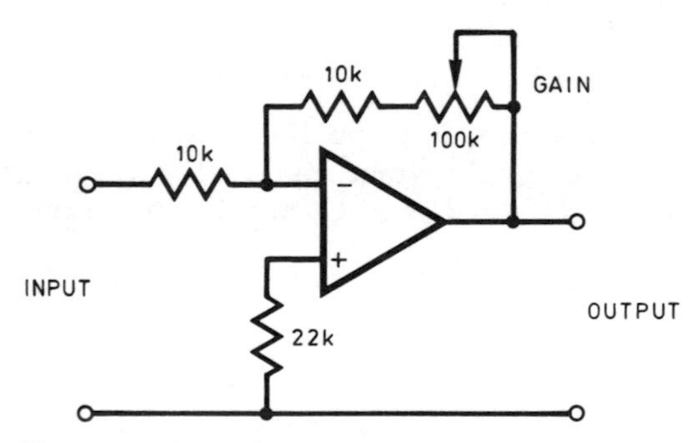

Figure 7.23 Variable gain amplifier (see Example 7.11)

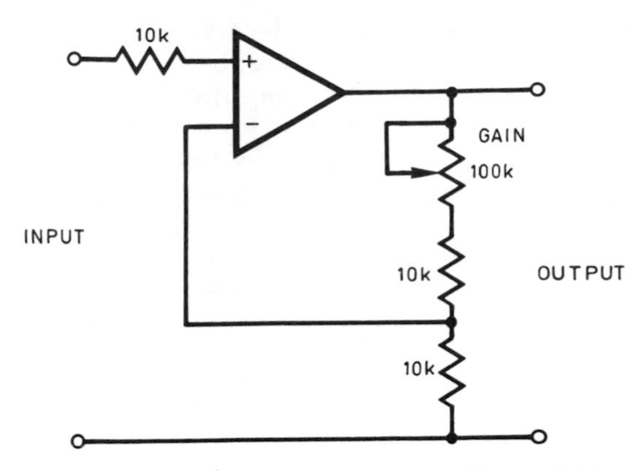

Figure 7.24 Non-inverting alternative to Figure 7.23

Example 7.12
Low-pass and high-pass filters are required with cut-off frequencies of 4kHz and 100Hz, respectively. Devise suitable circuit arrangements.

In the case of the low-pass filter, f_C = 4kHz and the circuit required is shown in Figure 7.21(a). The value of C is calculated from:

$$C = \frac{15915}{f_C} = \frac{15915}{4000} = 3.97\text{nF}$$ (realised by connecting a 27nF capacitor in series with a 4.7nF capacitor, see Appendix C)

In the case of the high-pass filter, f_C = 100Hz and the circuit required is shown in Figure 7.21(b). The value of C is calculated from:

$$C = \frac{15915}{f_C} = \frac{15915}{100} = 159.1\text{nF}$$ (realised by connecting a 270nF capacitor in series with a 390nF capacitor, see Appendix C)

Hints and tips
★ Operational amplifiers usually require positive and negative supply rail voltages in the region of ±9 to ±15V. For clarity, supply rails are often omitted from circuit diagrams.
★ Where only a single supply rail is available (e.g. +12V) it is usually possible to split the supply (using a resistive potential divider) to form a *half-supply rail*. This rail provides a reference input voltage (for either the inverting or non-inverting input depending upon the configuration employed). The half-supply rail

will require efficient decoupling (a 47μF or 100μF capacitor should be adequate) and a.c. signals will also require coupling capacitors at the input and output to remove the half-supply d.c. level.

★ The *gain × bandwidth product* is constant for a particular operational amplifier type. A trade-off thus exists between gain and bandwidth; the greater the gain the lower the bandwidth. Most modern operational amplifiers provide gain × bandwidth products in excess of 2MHz. *Slew rate* and gain × bandwidth product are closely related; an amplifier that offers a high slew rate will also provide a relatively large gain × bandwidth product.

Appendix A Abbreviations

a	anode
A	ampere
a.c.	alternating current
AC	alternating current
ACW	anti-clockwise
ADC	analogue-to-digital converter
a.f.	audio frequency
AF	audio frequency
a.g.c.	automatic gain control
AGC	automatic gain control
AlSb	aluminium antimonide
ALC	automatic level control
ALS	advanced low-power Schottky
ALSTTL	advanced low-power Schottky transistor-transistor logic
ALU	arithmetic logic unit
AM	amplitude modulation
ANSI	American National Standards Institute
ASA	American Standards Association
ASCII	American Standard Code for Information Interchange
ASIC	application specific integrated circuit
ATE	automatic test equipment
ATTD	avalanche transit time device
AUX	auxiliary
AWG	American Wire Gauge
b	base
BABT	British Approvals Board for Telecommunications

BARITT	barrier injected transit time diode
BCD	binary coded decimal
BCS	British Computer Society
BFO	beat frequency oscillator
BHJT	bipolar heterojunction transistor
BICMOS	bipolar complementary metal oxide semiconductor
BIMOS	bipolar metal oxide semiconductor
BJT	bipolar junction transistor
bps	bits per second
BSI	British Standards Institute

c	collector
C	capacitor
C	Coulomb
°C	degree Celsius
CAD	computer aided design
CB	citizens' band
CBIC	complementary bipolar integrated circuit
CCD	charge coupled device
CCW	counter-clockwise
cd	candela
Cd	cadmium
CDI	collector diffusion isolation
CdS	cadmium sulphide
CdTe	cadmium telluride
CERDIP	ceramic dual in-line package
CHINT	charge injection transistor
CLK	clock
cm	centimetre
CMOS	complementary metal oxide semiconductor
CMRR	common-mode rejection ratio
COMFET	conductivity modulated field effect transistor
c.p.s.	cycles per second
CPU	central processing unit
CRT	cathode ray tube
CRTC	cathode ray tube controller
c.s.a.	cross sectional area
CSA	Canadian Standards Association
c.t.	centre tap
C/T	current transformer
CTR	current transfer ratio
CTS	clear to send

CW	continuous wave
CW	counter-clockwise
d	drain
D	diode
D/A	digital-to-analogue
DAC	digital-to-analogue converter
dB	decibel
dBm	decibel (relative to 1mW)
DBS	direct broadcasting by satellite
dBV	decibel (relative to 1V)
d.c.	direct current
DC	direct current
DCE	data circuit-terminating equipment
DEMUX	demultiplexer
DIL	dual in-line
DIN	Deutsche Industrie Normal
DIP	dual in-line package
DMA	direct memory access
DMM	digital multimeter
DMOS	double diffused metal oxide semiconductor
DOS	disk operating system
DP	data processing
DPDT	double-pole, double-throw
DPST	double-pole, single-throw
DRAM	dynamic random access memory
DSB	double sideband
DSP	digital signal processing
DTE	data terminal equipment
DTL	diode-transistor logic
DUT	device under test
DVM	digital voltmeter
e	emitter
EAROM	electrically alterable read-only memory
EBCD	extended binary coded decimal
EBL	electron beam lithography
EBU	European Broadcasting Union
e.c.c.	earth continuity conductor
ECMA	European Computer Manufacturers' Association
EEPROM	electrically erasable programmable read-only memory
EEROM	electrically erasable read-only memory

E^2ROM	electrically erasable read-only memory
ECL	electrically coupled logic
EHT	extra high tension
EIA	Electronic Industries Association
EITB	Engineering Industry Training Board
ELED	edge-emitting light emitting diode
EM	electromagnetic
EMC	electromagnetic compatibility
e.m.f.	electromotive force
EMI	electromagnetic interference
EMP	electromagnetic pulse
EPROM	erasable-programmable read-only memory
EROM	erasable read-only memory
ESD	electrostatic discharge
f	frequency
F	Farad
FAX	facsimile
f_c	cut-off frequency
FDC	floppy disk controller
FDM	frequency division multiplexing
FET	field effect transistor
FIFO	first-in first-out
FLL	frequency-locked loop
FM	frequency modulation
FPGA	field programmable gate array
FPLA	field programmable logic array
FPROM	field programmable read-only memory
FRU	field replaceable unit
FSD	full-scale deflection
FSK	frequency-shift keying
FSM	frequency shift modulation
FSTV	fast-scan television
f_T	transition frequency
g	gate
G	giga ($\times 10^9$)
GaAlAS	gallium aluminium arsenide
GaAs	gallium arsenide
GaInAsP	gallium indium arsenide phosphide
GaP	gallium phosphide
GDP	graphic display processor
Ge	germanium

GHz	gigahertz
GND	ground
GPIB	general purpose interface bus
GTO	gate turn off
H	henry
HDTV	high definition television
hex	hexadecimal
HEMT	high electron mobility transistor
HF	high frequency
HMOS	high-speed metal oxide semiconductor
HPIB	Hewlett-Packard interface bus
Hz	hertz
IA5	International Alphabet No. 5
i.c.	integrated circuit
IC	integrated circuit
ICE	in-circuit emulation
IEC	International Electrotechnical Commission
IEE	Institution of Electrical Engineers
IEEE	Institute of Electrical and Electronic Engineers
IEEIE	Institution of Electrical and Electronics Incoporated Engineers
IERE	Institution of Electronic and Radio Engineers
IF	intermediate frequency
IGFET	insulated gate field effect transistor
IGT	insulated gate transistor
IKBS	intelligent knowledge-based systems
I^2L	integrated injection logic
IMD	intermodulation
IMPATT	impact ionization avalanche transit time diode
In	indium
InP	indium phosphide
InSb	indium antimonide
I/O	input/output
I/P	input
IR	infra-red
ISL	integrated Schottky logic
ISO	International Standards Organisation
IT	information technology
ITeC	Information Technology Centre
ITU	International Telecommunications Union

J	joule
JAN	joint Army/Navy
JFET	junction field effect transistor
JUGFET	junction gate field effect transistor
k	cathode
k	kilo ($\times 10^3$)
K	binary kilo ($\times 1024$)
°K	degree Kelvin
Kb	kilobyte
kg	kilogramme
kHz	kilohertz
km	kilometre
kΩ	kilohm
kV	kilovolt
L	inductor
LAN	local area network
LCC	leadless chip carrier
LCD	liquid crystal display
LDR	light dependent resistor
LED	light emitting diode
LF	low frequency
LIFO	last-in first-out
lm	lumen
LOCMOS	locally oxidised complementary metal oxide semiconductor
LPE	liquid phase epitaxy
LS	low-power Schottky
LSA	limited space-charge accumulation
LSB	lower sideband
LSB	least-significant bit
LSD	least-significant digit
LSI	large scale integration
LVRT	linear variable reluctance transducer
LVDT	linear variable differential transducer
lx	lux
m	metre
m	milli ($\times 10^{-3}$)
M	mega ($\times 10^6$)
mA	milliampere

MAC	multiplexed analogue components
MAP	Manufacturing Automation Protocol
MBE	molecular beam epitaxy
MESFET	metal semiconductor field effect transistor
MF	medium frequency
MFM	modified frequency modulation
mH	millihenry
MHz	megahertz
MINIDIP	miniature dual-in-line package
MISFET	metal insulator silicon field effect transistor
mm	millimetre
MMIC	monolithic microwave integrated circuit
MΩ	megohm
MOCVD	metal organic chemical vapour deposition
MODEM	modulator/demodulator
MOS	metal oxide silicon
MOSFET	metal oxide silicon field effect transistor
MOST	metal oxide semiconductor transistor
MPU	microprocessor unit
MPX	multiplex
MSB	most-significant bit
MSD	most-significant digit
MSI	medium scale integration
mt	main terminal
MTBF	mean time between failure
MTTR	mean time to repair
MTTF	mean time to failure
MUX	multiplexer
mV	millivolt
mW	milliwatt
n	nano ($\times 10^{-9}$)
n	negative
N	negative
N	Newton
n.c.	not connected
NERFET	negative resistance field effect transistor
nF	nanofarad
nH	nanohenry
Ni	nickel
NiCd	nickel cadmium
NMOS	N-channel metal oxide semiconductor
NOVRAM	non-volatile random access memory

n.p.n.	negative-positive-negative
NPN	negative-positive-negative
NRFET	negative resistance field effect transistor
NRZ	non-return to zero
n.t.c.	negative temperature coefficient
NTC	negative temperature coefficient
NTSC	National Television Systems Committee
nV	nanovolt
nW	nanowatt
Ω	ohm
o.c.	open circuit
OCR	optical character recognition
OEM	original equipment manufacturer
OIC	optically coupled integrated circuit
O/P	output
OSI	open systems interconnection
p	pico ($\times 10^{-12}$)
p	positive
P	positive
PAL	phase alternate line
PAL	programmable array logic
PAM	pulse amplitude modulation
PC	personal computer
p.c.b.	printed circuit board
PCB	printed circuit board
PCM	pulse code modulation
p.d.	potential difference
pF	picofarad
PF	power factor
PIO	parallel input/output
PIN	positive intrinsic negative
PINFET	positive intrinsic negative field effect transistor
PIPO	parallel input, parallel output
PISO	parallel input, serial output
pk	peak
pk-pk	peak-to-peak
PL	plug
PLA	programmable logic array
PLC	programmable logic controller
PLL	phase-locked loop
PM	pulse modulation

PML	programmable macro logic
PMOS	P-channel metal oxide semiconductor
PMR	private mobile radio
p.n.	positive-negative
PN	positive-negative
p.n.p.	positive-negative-positive
PNP	positive-negative-positive
PPI	programmable parallel interface
ppm	parts per million
PPM	pulse position modulation
pps	pulses per second
prf	pulse repetition frequency
PROM	programmable read-only memory
PSD	phase sensitive detector
PSE	packet switching exchange
PSK	phase-shift keying
PSRAM	pseudo-static random access memory
PSU	power supply unit
p.t.c.	positive temperature coefficient
PTC	positive temperature coefficient
PTFE	polytetrafluoroethylene
PUJT	programmable unijunction transistor
PUT	programmable unijunction transistor
PVC	polyvinylchloride
PWB	printed wiring board
PWM	pulse width modulation
Q	quality factor
Q	transistor
QAM	quadrature amplitude modulation
QIL	quad in-line
R	resistor
rad	radian
RAM	random access memory
RB	ripple blanking
r.f.	radio frequency
RFI	radio frequency interference
RGB	red, green, blue
RIAA	Radio Industry Association of America
RISC	reduced instruction set code
RMM	read-mostly memory
r.m.s.	root mean square

RMS	root mean square
ROM	read-only memory
RTC	real-time clock
RTL	resistor-transistor logic
RTS	request to send
RTTY	radio teletype
RX	receive
RZ	return to zero
s	source
s	second
S	Siemen
S	switch
SBC	single-board computer
s.c.	short circuit
SCR	silicon controlled rectifier
SCRAM	static column dynamic random access memory
SCS	silicon controlled switch
Se	selenium
SECAM	Sequential Couleur à Mémoire
Si	silicon
SI	System International
SIL	single in-line
SIO	serial input/output
SIPO	serial input, parallel output
SISO	serial input, serial output
SK	socket
SLED	surface-emitting light emitting diode
SLSI	super large scale integration
SMC	surface mounted component
SMD	surface mounted device
SME	Society of Manufacturing Engineers
SMT	surface mount technology
S/N	signal-to-noise
SNA	systems network architecture
SNR	signal-to-noise ratio
SOS	silicon on sapphire
SPDT	single-pole, double-throw
SPST	single-pole, single-throw
sr	steradian
SRAM	static random access memory
SSB	single sideband
SSI	small scale integration

SSTV	slow-scan television
SWG	standard wire gauge
SWR	standing wave ratio
t	time
T	Tesla
T	transformer
TDM	time division multiplexing
TED	transferred electron device
THD	total harmonic distortion
TOP	technical and office protocol
tot	total
TP	test point
TR	transformer
TRAPATT	trapped plasma avalanche triggered diode
TTL	transistor-transistor logic
TTY	teletype
TV	television
TX	transmit
μ	micro
μA	microampere
UART	universal asynchronous receiver/transmitter
μF	microfarad
μH	microhenry
UHF	ultra high frequency
UJT	unijunction transistor
ULA	uncommitted logic array
ULSI	ultra large scale integration
μm	micrometre
UMOS	U-channel metal oxide semiconductor
UPS	uninterruptible power supply
USART	universal synchronous/asynchronous receiver/ transmitter
USB	upper sideband
μV	microvolt
UV	ultra-violet
μW	microwatt
V	volt
VAN	value added network
VDR	voltage dependent resistor
VDU	visual display unit

VFC	voltage-to-frequency converter
VFO	variable frequency oscillator
VHF	very high frequency
VLSI	very large scale integration
VMOS	V-channel metal oxide semiconductor
VPE	vapour phase epitaxy
VRAM	video random access memory
VSB	vestigial sideband
VU	volume units
WAN	wide area network
W	watt
Wb	weber
WORM	write once read many
X	crystal
X	reactance
Y	crystal
Z	impedance

Appendix B
Symbols for electronic devices

A number of standard graphical symbols are used to represent electronic components within circuit diagrams. These symbols are defined in BS3939 in the UK, and ANSI Y32.2 in the USA. The differences which exist between the two standards are thankfully relatively minor and do not generally give any cause for concern. Furthermore, 'in house' standards are often adopted by equipment manufacturers and publishers and these may differ from the accepted national standards.

The symbols depicted in this appendix form a subset of the most commonly used graphical symbols for electronic components. It is particularly important to note that the symbols used for logic gates are those taken from the more commonly used MIL/ANSI standard and that these differ markedly from those defined in the equivalent British Standard.

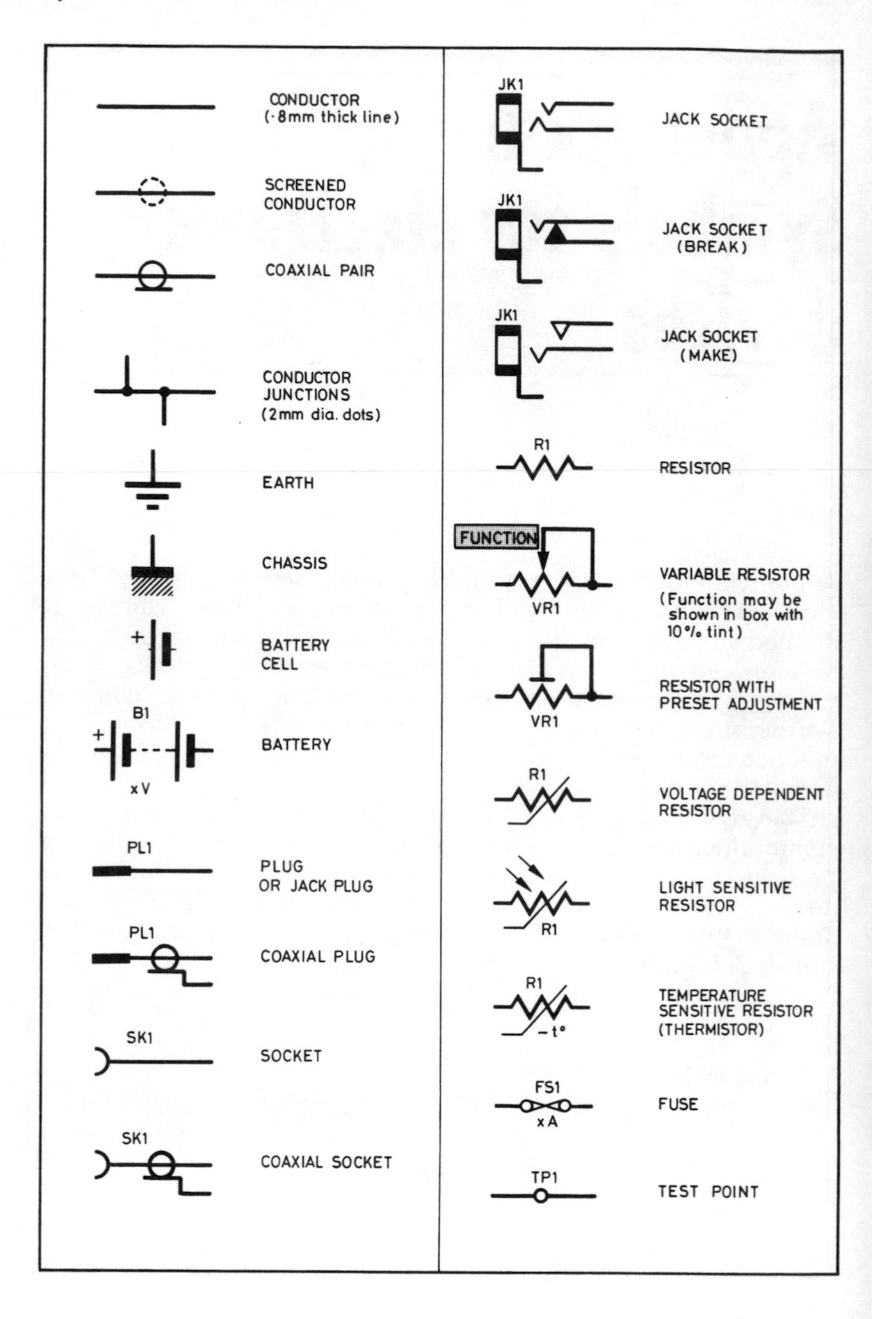
CONDUCTOR
(·8mm thick line)
SCREENED
CONDUCTOR
COAXIAL PAIR
CONDUCTOR
JUNCTIONS
(2mm dia. dots)
EARTH
CHASSIS
+
BATTERY
CELL
B1
+
xV
BATTERY
PL1
PLUG
OR JACK PLUG
PL1
COAXIAL PLUG
SK1
SOCKET
SK1
COAXIAL SOCKET
JK1
JACK SOCKET
JK1
JACK SOCKET
(BREAK)
JK1
JACK SOCKET
(MAKE)
R1
RESISTOR
FUNCTION
VR1
VARIABLE RESISTOR
(Function may be
shown in box with
10% tint)
VR1
RESISTOR WITH
PRESET ADJUSTMENT
R1
VOLTAGE DEPENDENT
RESISTOR
R1
LIGHT SENSITIVE
RESISTOR
R1
−t°
TEMPERATURE
SENSITIVE RESISTOR
(THERMISTOR)
FS1
xA
FUSE
TP1
TEST POINT

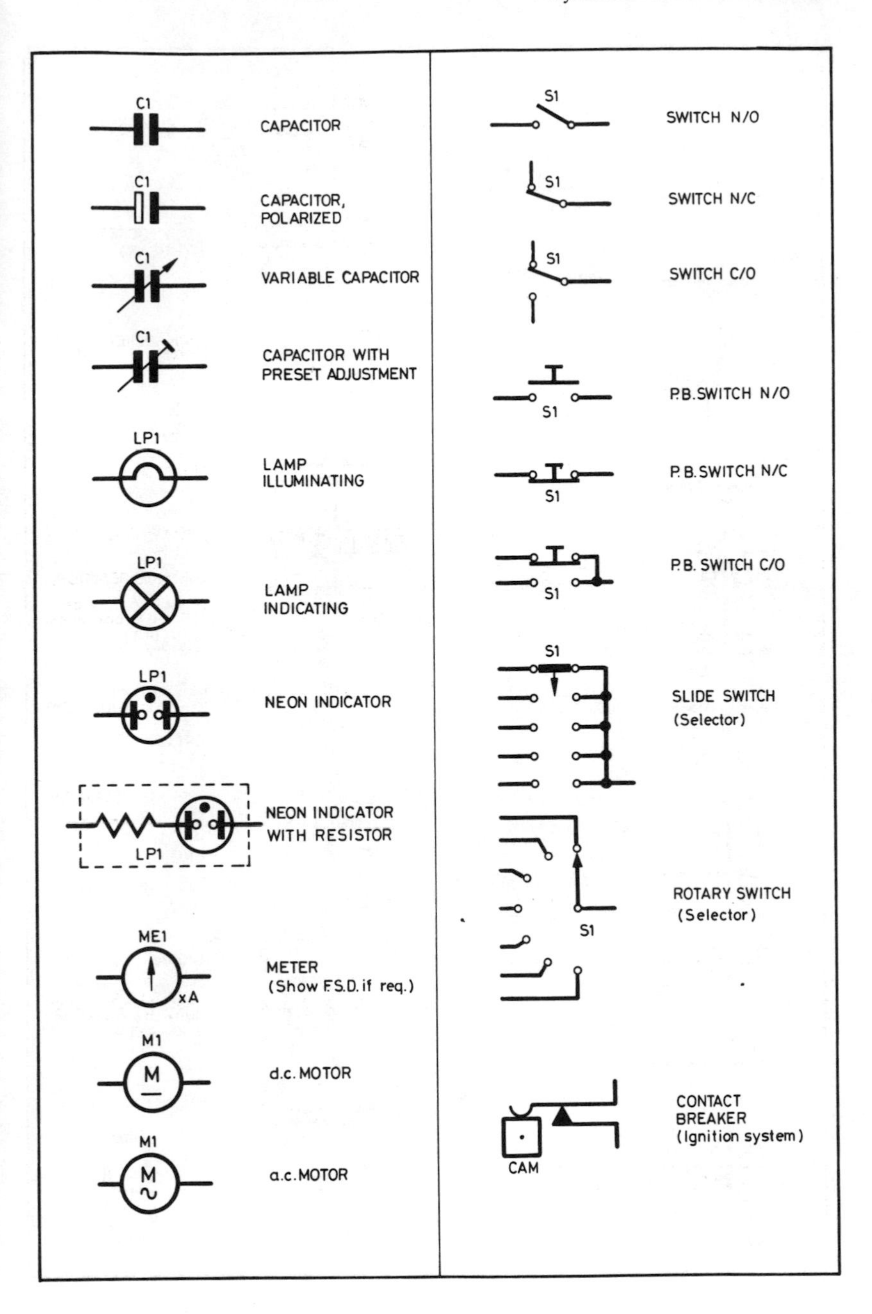
C1
CAPACITOR
C1
CAPACITOR, POLARIZED
C1
VARIABLE CAPACITOR
C1
CAPACITOR WITH PRESET ADJUSTMENT
LP1
LAMP ILLUMINATING
LP1
LAMP INDICATING
LP1
NEON INDICATOR
LP1
NEON INDICATOR WITH RESISTOR
ME1
xA
METER (Show F.S.D. if req.)
M1
M
d.c. MOTOR
M1
M
a.c. MOTOR
S1
SWITCH N/O
S1
SWITCH N/C
S1
SWITCH C/O
S1
P.B. SWITCH N/O
S1
P.B. SWITCH N/C
S1
P.B. SWITCH C/O
S1
SLIDE SWITCH (Selector)
S1
ROTARY SWITCH (Selector)
CAM
CONTACT BREAKER (Ignition system)

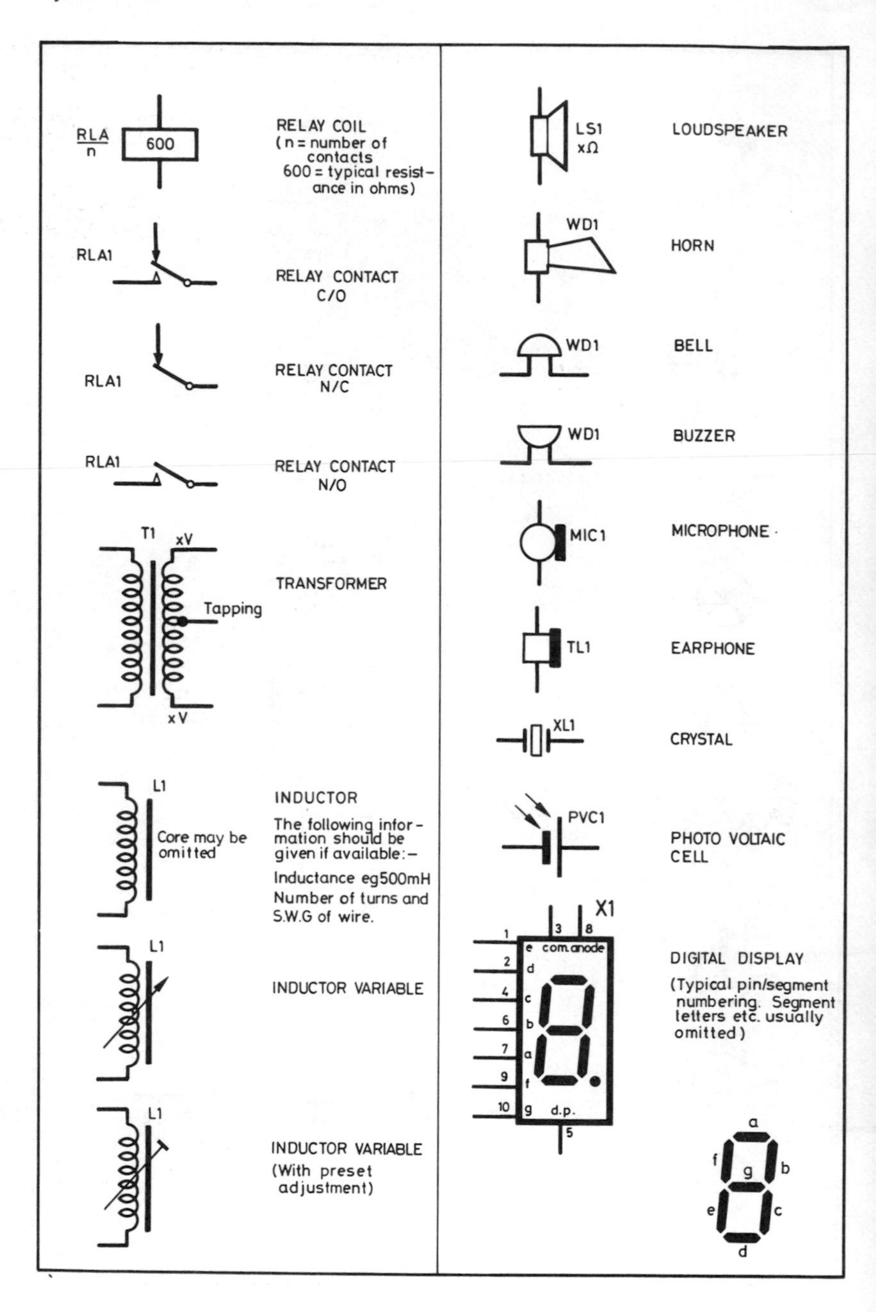
RLA
n
600
RELAY COIL
(n = number of contacts
600 = typical resistance in ohms)
RLA1
RELAY CONTACT
C/O
RLA1
RELAY CONTACT
N/C
RLA1
RELAY CONTACT
N/O
T1
xV
Tapping
TRANSFORMER
xV
L1
Core may be omitted
INDUCTOR
The following information should be given if available:–
Inductance eg500mH
Number of turns and S.W.G of wire.
L1
INDUCTOR VARIABLE
L1
INDUCTOR VARIABLE
(With preset adjustment)
LS1
xΩ
LOUDSPEAKER
WD1
HORN
WD1
BELL
WD1
BUZZER
MIC1
MICROPHONE
TL1
EARPHONE
XL1
CRYSTAL
PVC1
PHOTO VOLTAIC CELL
X1
3
8
1
e
com.anode
2
d
4
c
6
b
7
a
9
f
10
g
d.p.
5
DIGITAL DISPLAY
(Typical pin/segment numbering. Segment letters etc. usually omitted)
a
f
g
b
e
c
d

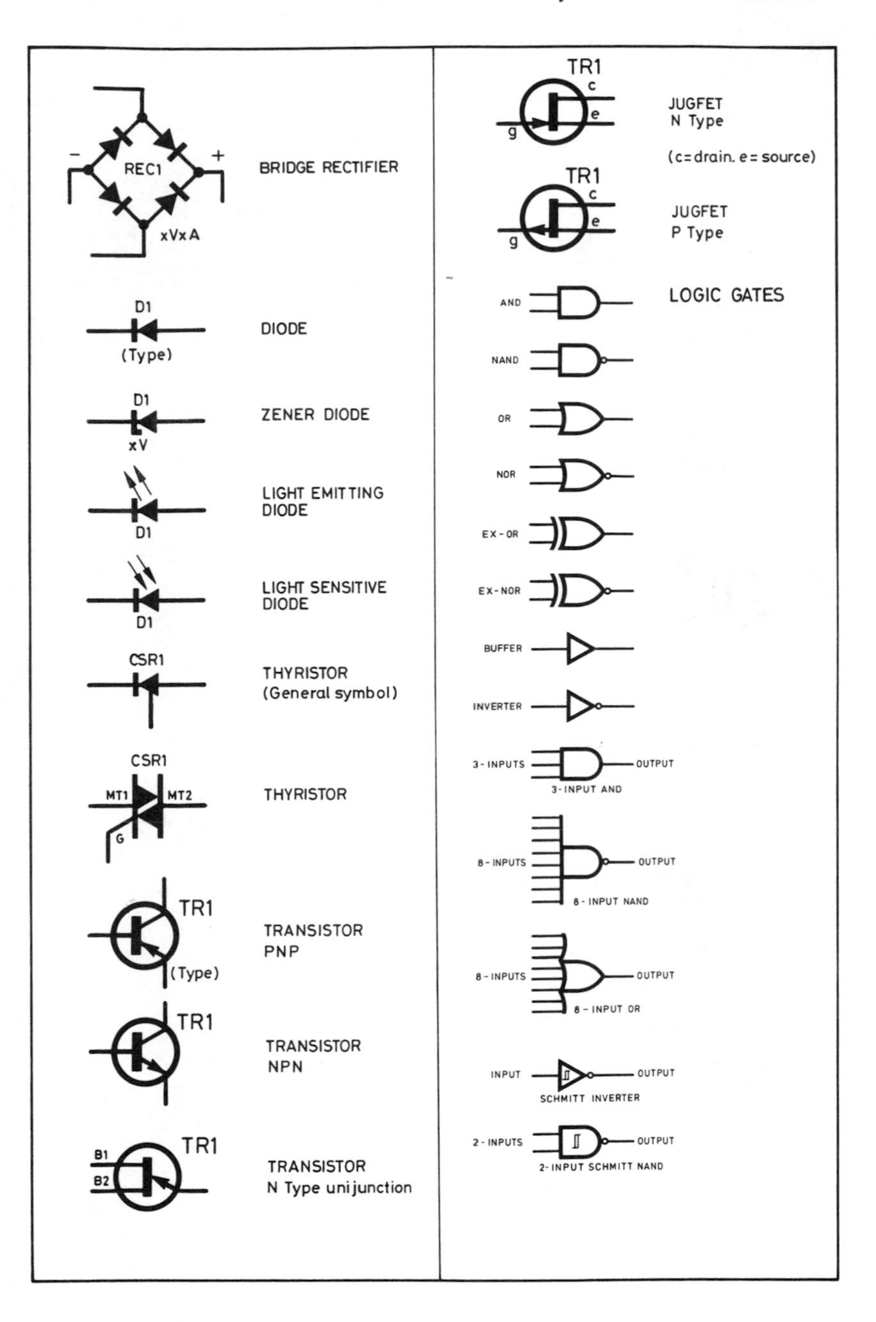
REC1
xVxA
BRIDGE RECTIFIER
D1
(Type)
DIODE
D1
xV
ZENER DIODE
D1
LIGHT EMITTING DIODE
D1
LIGHT SENSITIVE DIODE
CSR1
THYRISTOR (General symbol)
CSR1
MT1
MT2
G
THYRISTOR
TR1
(Type)
TRANSISTOR PNP
TR1
TRANSISTOR NPN
B1
B2
TR1
TRANSISTOR N Type unijunction
TR1
c
e
g
JUGFET N Type
(c=drain. e=source)
TR1
c
e
g
JUGFET P Type
LOGIC GATES
AND
NAND
OR
NOR
EX-OR
EX-NOR
BUFFER
INVERTER
3-INPUTS
OUTPUT
3-INPUT AND
8-INPUTS
OUTPUT
8-INPUT NAND
8-INPUTS
OUTPUT
8-INPUT OR
INPUT
OUTPUT
SCHMITT INVERTER
2-INPUTS
OUTPUT
2-INPUT SCHMITT NAND

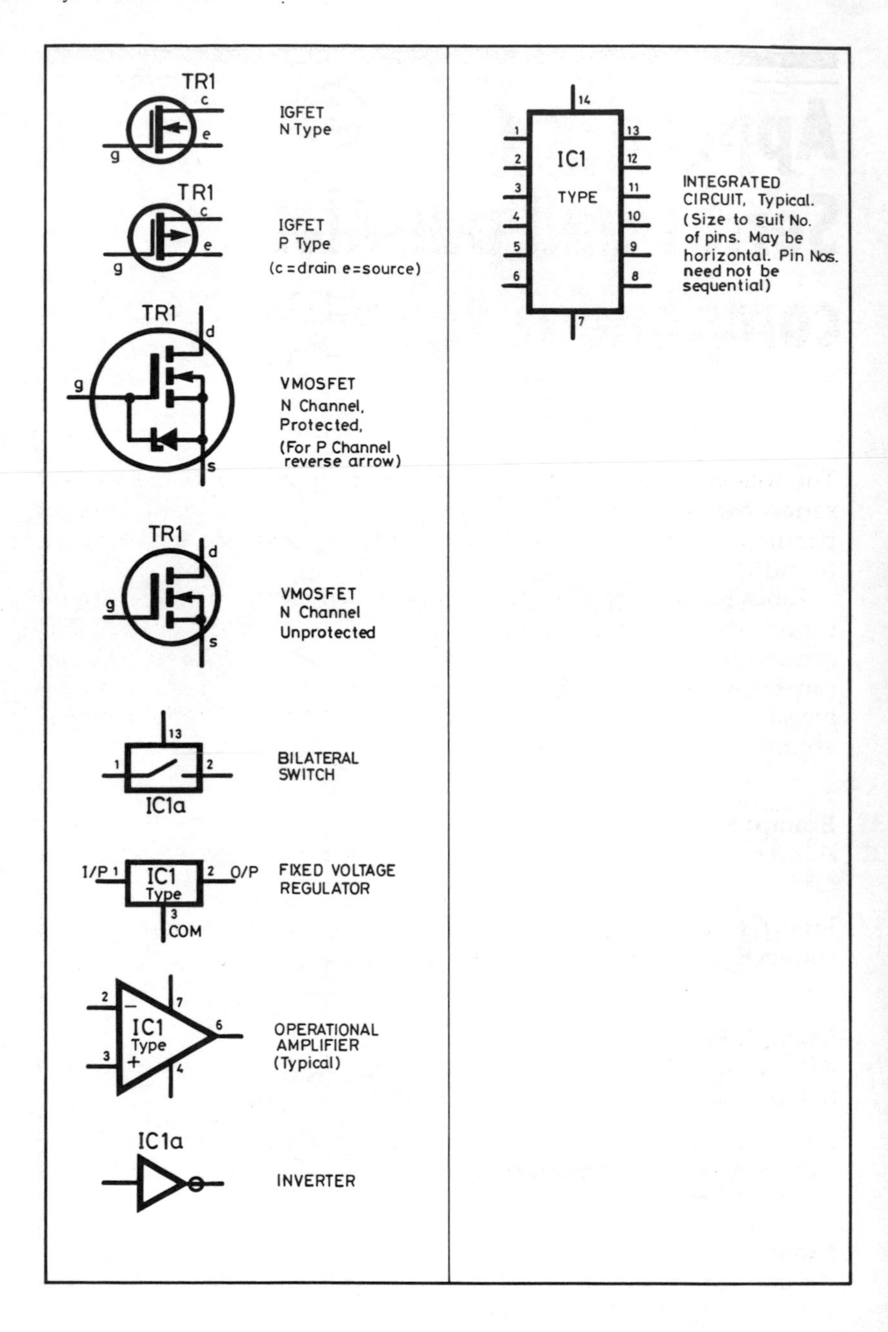
TR1
c
e
g
IGFET
N Type
TR1
c
e
g
IGFET
P Type
(c =drain e=source)
TR1
d
g
s
VMOSFET
N Channel,
Protected,
(For P Channel
reverse arrow)
TR1
d
g
s
VMOSFET
N Channel
Unprotected
13
1
2
IC1a
BILATERAL
SWITCH
I/P 1
IC1
Type
2 O/P
3
COM
FIXED VOLTAGE
REGULATOR
2
7
IC1
Type
6
3
4
OPERATIONAL
AMPLIFIER
(Typical)
IC1a
INVERTER
14
1
2
3
4
5
6
13
12
11
10
9
8
7
IC1
TYPE
INTEGRATED
CIRCUIT, Typical.
(Size to suit No.
of pins. May be
horizontal. Pin Nos.
need not be
sequential)

Appendix C
Series and parallel component tables

The following tables can be used to determine the values of E-12 series components which, when connected in either series or parallel, will provide a particular value of resistance, capacitance, or inductance.

Table 1 is for a resistors or inductors connected in series and for capacitors connected in parallel. Table 2 is for resistors or inductors connected in parallel and for capacitors connected in series. Values can be scaled up or down (by manipulating the decimal point) *provided* that all values (columns, rows and values at intersections) are treated in the same way.

Example C.1
A resistance value of 25Ω is required. What two values of resistor connected in series will produce this value?

From Table 1, a value of 25Ω would result from a value of 10Ω connected in series with a value of 15Ω.

Example C.2
An inductance value of 90μH is required. What two values of inductor connected in series will produce this value?

From Table 1, a value of 90.0μH would result from a (column) value of 22μH connected in series with a (row) value of 68μH.

Example C.3
An inductance value of 40mH is required. What two values of inductor connected in parallel will produce this value?

Table 1 Resistors or inductors connected in series. Capacitors connected in parallel

	First value (column)											
	10	12	15	18	22	27	33	39	47	56	68	82
10	20	22	25	28	32	37	43	49	57	66	78	92
12	22	24	27	30	34	39	45	51	59	68	80	94
15	25	27	30	33	37	42	48	54	62	71	83	97
18	28	30	33	36	40	45	51	57	65	74	86	100
22	32	34	37	40	44	49	55	61	69	78	90	104
27	37	39	42	45	49	54	60	66	74	83	95	109
33	43	45	48	51	55	60	66	72	80	89	101	115
39	49	51	54	57	61	66	72	78	86	95	107	121
47	57	59	62	65	69	74	80	86	94	103	115	129
56	66	68	71	74	78	83	89	95	103	112	124	138
68	78	80	83	86	90	95	101	107	115	124	136	150
82	92	94	97	100	104	109	115	121	129	138	150	164
100	110	112	115	118	122	127	133	139	147	156	168	182
120	130	132	135	138	142	147	153	159	167	176	188	202
150	160	162	165	168	172	177	183	189	197	206	218	232
180	190	192	195	198	202	207	213	219	227	236	248	262
220	230	232	235	238	242	247	253	259	267	276	288	302
270	280	282	285	288	292	297	303	309	317	326	338	352
330	340	342	345	348	352	357	363	369	377	386	398	412
390	400	402	405	408	412	417	423	429	437	446	458	472
470	480	482	485	488	492	497	503	509	517	526	538	552
560	570	572	575	578	582	587	593	599	607	616	628	642
680	690	692	695	698	702	707	713	719	727	736	748	762
820	830	832	835	838	842	847	853	859	867	876	888	902
1000	1010	1012	1015	1018	1022	1027	1033	1039	1047	1056	1068	1082

Second value (row)

From Table 2, a value of 40.0mH would result from a (column) value of 47mH connected in parallel with a (row) value of 270mH.

Example C.4

A capacitance value of 6μF is required. What two values of capacitor connected in parallel will produce this value?

From Table 1, a value of 60μF would result from a (column) value of 33μF connected in parallel with a (row) value of 27μF. These values can all be scaled down by a factor of ten by simply shifting the decimal points one space to the left (i.e. dividing each by ten). The required values would thus be 3.3μF and 2.7μF connected in parallel.

Table 2 Resistors or inductors connected in parallel. Capacitors connected in series

First value (column)

	10	12	15	18	22	27	33	39	47	56	68	82
10	5.0	5.5	6.0	6.4	6.9	7.3	7.7	8.0	8.2	8.5	8.7	8.9
12	5.5	6.0	6.7	7.2	7.8	8.3	8.8	9.2	9.6	9.9	10.2	10.5
15	6.0	6.7	7.5	8.2	8.9	9.6	10.3	10.8	11.4	11.8	12.3	12.7
18	6.4	7.2	8.2	9.0	9.9	10.8	11.6	12.3	13.0	13.6	14.2	14.8
22	6.9	7.8	8.9	9.9	11.0	12.1	13.2	14.1	15.0	15.8	16.6	17.3
27	7.3	8.3	9.6	10.8	12.1	13.5	14.9	16.0	17.1	18.2	19.3	20.3
33	7.7	8.8	10.3	11.6	13.2	14.9	16.5	17.9	19.4	20.8	22.2	23.5
39	8.0	9.2	10.8	12.3	14.1	16.0	17.9	19.5	21.3	23.0	24.8	26.4
47	8.2	9.6	11.4	13.0	15.0	17.1	19.4	21.3	23.5	25.6	27.8	29.9
56	8.5	9.9	11.8	13.6	15.8	18.2	20.8	23.0	25.6	28.0	30.7	33.3
68	8.7	10.2	12.3	14.2	16.6	19.3	22.2	24.8	27.8	30.7	34.0	37.2
82	8.9	10.5	12.7	14.8	17.3	20.3	23.5	26.4	29.9	33.3	37.2	41.0
100	9.1	10.7	13.0	15.3	18.0	21.3	24.8	28.1	32.0	35.9	40.5	45.1
120	9.2	10.9	13.3	15.7	18.6	22.0	25.9	29.4	33.8	38.2	43.4	48.7
150	9.4	11.1	13.6	16.1	19.2	22.9	27.0	31.0	35.8	40.8	46.8	53.0
180	9.5	11.3	13.8	16.4	19.6	23.5	27.9	32.1	37.3	42.7	49.4	56.3
220	9.6	11.4	14.0	16.6	20.0	24.0	28.7	33.1	38.7	44.6	51.9	59.7
270	9.6	11.5	14.2	16.9	20.3	24.5	29.4	34.1	40.0	46.4	54.3	62.9
330	9.7	11.6	14.3	17.1	20.6	25.0	30.0	34.9	41.1	47.9	56.4	65.7
390	9.8	11.6	14.4	17.2	20.8	25.3	30.4	35.5	41.9	49.0	57.9	67.8
470	9.8	11.7	14.5	17.3	21.0	25.5	30.8	36.0	42.7	50.0	59.4	69.8
560	9.8	11.7	14.6	17.4	21.2	25.8	31.2	36.5	43.4	50.9	60.6	71.5
680	9.9	11.8	14.7	17.5	21.3	26.0	31.5	36.9	44.0	51.7	61.8	73.2
820	9.9	11.8	14.7	17.6	21.4	26.1	31.7	37.2	44.5	52.4	62.8	74.5
1000	9.9	11.9	14.8	17.7	21.5	26.3	31.9	37.5	44.9	53.0	63.7	75.8

Second value (row)

Appendix D Decibels and power, voltage and current ratios

dB	*Power ratio*	*Voltage/current ratio*
−99	1.258925×10^{-10}	1.122018×10^{-5}
−98	1.584893×10^{-10}	1.258925×10^{-5}
−97	1.995262×10^{-10}	1.412538×10^{-5}
−96	2.511887×10^{-10}	1.584893×10^{-5}
−95	3.162278×10^{-10}	1.778279×10^{-5}
−94	3.981072×10^{-10}	1.995262×10^{-5}
−93	5.011873×10^{-10}	2.238721×10^{-5}
−92	6.309573×10^{-10}	2.511886×10^{-5}
−91	7.943282×10^{-10}	2.818383×10^{-5}
−90	1.000000×10^{-9}	3.162278×10^{-5}
−89	1.258925×10^{-9}	3.548134×10^{-5}
−88	1.584893×10^{-9}	3.981072×10^{-5}
−87	1.995262×10^{-9}	4.466836×10^{-5}
−86	2.511886×10^{-9}	5.011872×10^{-5}
−85	3.162278×10^{-9}	5.623413×10^{-5}
−84	3.981072×10^{-9}	6.309574×10^{-5}
−83	5.011872×10^{-9}	7.079458×10^{-5}
−82	6.309573×10^{-9}	7.943282×10^{-5}
−81	7.943282×10^{-9}	8.912510×10^{-5}
−80	1.000000×10^{-8}	1.000000×10^{-4}
−79	1.258925×10^{-8}	1.122018×10^{-4}
−78	1.584893×10^{-8}	1.258925×10^{-4}
−77	1.995262×10^{-8}	1.412538×10^{-4}
−76	2.511887×10^{-8}	1.584893×10^{-4}
−75	3.162278×10^{-8}	1.778279×10^{-4}
−74	3.981072×10^{-8}	1.995262×10^{-4}
−73	5.011873×10^{-8}	2.238721×10^{-4}
−72	6.309573×10^{-8}	2.511886×10^{-4}
−71	7.943282×10^{-8}	2.818383×10^{-4}
−70	1.000000×10^{-7}	3.162278×10^{-4}
−69	1.258925×10^{-7}	3.548134×10^{-4}
−68	1.584893×10^{-7}	3.981072×10^{-4}
−67	1.995262×10^{-7}	4.466836×10^{-4}
−66	2.511887×10^{-7}	5.011872×10^{-4}
−65	3.162278×10^{-7}	5.623413×10^{-4}

dB	*Power ratio*	*Voltage/current ratio*
−64	3.981072×10^{-7}	6.309574×10^{-4}
−63	5.011872×10^{-7}	7.079458×10^{-4}
−62	6.309573×10^{-7}	7.943282×10^{-4}
−61	7.943282×10^{-7}	8.912510×10^{-4}
−60	1.000000×10^{-6}	1.000000×10^{-3}
−59	1.258925×10^{-6}	1.122018×10^{-3}
−58	1.584893×10^{-6}	1.258925×10^{-3}
−57	1.995262×10^{-6}	1.412538×10^{-3}
−56	2.511886×10^{-6}	1.584893×10^{-3}
−55	3.162278×10^{-6}	1.778279×10^{-3}
−54	3.981072×10^{-6}	1.995262×10^{-3}
−53	5.011872×10^{-6}	2.238721×10^{-3}
−52	6.309574×10^{-6}	2.511886×10^{-3}
−51	7.943282×10^{-6}	2.818383×10^{-3}
−50	1.000000×10^{-5}	3.162278×10^{-3}
−49	1.258925×10^{-5}	3.548134×10^{-3}
−48	1.584893×10^{-5}	3.981072×10^{-3}
−47	1.995262×10^{-5}	4.466836×10^{-3}
−46	2.511886×10^{-5}	5.011872×10^{-3}
−45	3.162278×10^{-5}	5.623413×10^{-3}
−44	3.981072×10^{-5}	6.309574×10^{-3}
−43	5.011872×10^{-5}	7.079458×10^{-3}
−42	6.309574×10^{-5}	7.943282×10^{-3}
−41	7.943282×10^{-5}	8.912509×10^{-3}
−40	1.000000×10^{-4}	0.01000000
−39	1.258925×10^{-4}	0.01122018
−38	1.584893×10^{-4}	0.01258925
−37	1.995262×10^{-4}	0.01412538
−36	2.511886×10^{-4}	0.01584893
−35	3.162278×10^{-4}	0.01778279
−34	3.981072×10^{-4}	0.01995262
−33	5.011872×10^{-4}	0.02238721
−32	6.309574×10^{-4}	0.02511887
−31	7.943282×10^{-4}	0.02818383
−30	1.000000×10^{-3}	0.03162277
−29	1.258925×10^{-3}	0.03548134
−28	1.584893×10^{-3}	0.03981072
−27	1.995262×10^{-3}	0.04466836
−26	2.511886×10^{-3}	0.05011872
−25	3.162278×10^{-3}	0.05623413
−24	3.981072×10^{-3}	0.06309573
−23	5.011872×10^{-3}	0.07079457
−22	6.309574×10^{-3}	0.07943282
−21	7.943282×10^{-3}	0.08912510
−20	0.01000000	0.1000000
−19	0.01258925	0.1122018
−18	0.01584893	0.1258925
−17	0.01995262	0.1412538
−16	0.02511887	0.1584893
−15	0.03162277	0.1778279
−14	0.03981072	0.1995262
−13	0.05011872	0.2238721
−12	0.06309573	0.2511886
−11	0.07943282	0.2818383
−10	0.1000000	0.3162278
−9	0.1258925	0.3548134
−8	0.1584893	0.3981072

Decibels and power, voltage and current ratios

dB	*Power ratio*	*Voltage/current ratio*
−7	0.1995262	0.4466836
−6	0.2511886	0.5011872
−5	0.3162278	0.5623413
−4	0.3981072	0.6309574
−3	0.5011872	0.7079458
−2	0.6309574	0.7943282
−1	0.7943282	0.8912510
0	1.0000000	1.0000000
1	1.258925	1.122018
2	1.584893	1.258925
3	1.995262	1.412538
4	2.511886	1.584893
5	3.162278	1.778279
6	3.981072	1.995262
7	5.011872	2.238721
8	6.309574	2.511886
9	7.943282	2.818383
10	10.00000	3.162278
11	12.58925	3.548134
12	15.84893	3.981072
13	19.95262	4.466836
14	25.11886	5.011872
15	31.62278	5.623413
16	39.81072	6.309574
17	50.11872	7.079458
18	63.09573	7.943282
19	79.43282	8.912509
20	100.0000	10.00000
21	125.8925	11.22018
22	158.4893	12.58925
23	199.5262	14.12538
24	251.1886	15.84893
25	316.2278	17.78279
26	398.1072	19.95262
27	501.1872	22.38721
28	630.9573	25.11886
29	794.3282	28.18383
30	1000.000	31.62278
31	1258.925	35.48134
32	1584.893	39.81072
33	1995.262	44.66836
34	2511.886	50.11872
35	3162.278	56.23413
36	3981.072	63.09573
37	5011.873	70.79458
38	6309.573	79.43282
39	7943.282	89.12509
40	10000.00	100.0000
41	12589.25	112.2018
42	15848.93	125.8925
43	19952.62	141.2538
44	25118.86	158.4893
45	31622.78	177.8279
46	39810.72	199.5262
47	50118.72	223.8721
48	63095.73	251.1886
49	79432.82	281.8383

dB	*Power ratio*	*Voltage/current ratio*
50	100000.0	316.2278
51	125892.5	354.8134
52	158489.3	398.1072
53	199526.2	446.6836
54	251188.6	501.1872
55	316227.8	562.3413
56	398107.2	630.9573
57	501187.2	707.9458
58	630957.3	794.3282
59	794328.2	891.2509
60	1000000	1000.000
61	1258925	1122.018
62	1584893	1258.925
63	1995262	1412.538
64	2511886	1584.893
65	3162278	1778.279
66	3981072	1995.262
67	5011872	2238.721
68	6309573	2511.886
69	7943282	2818.383
70	1.000000×10^7	3162.278
71	1.258925×10^7	3548.134
72	1.584893×10^7	3981.072
73	1.995262×10^7	4466.836
74	2.511886×10^7	5011.873
75	3.162278×10^7	5623.413
76	3.981072×10^7	6309.573
77	5.011872×10^7	7079.458
78	6.309574×10^7	7943.282
79	7.943282×10^7	8912.51
80	1.000000×10^8	10000.00
81	1.258925×10^8	11220.18
82	1.584893×10^8	12589.25
83	1.995262×10^8	14125.37
84	2.511886×10^8	15848.93
85	3.162278×10^8	17782.79
86	3.981072×10^8	19952.62
87	5.011872×10^8	22387.21
88	6.309573×10^8	25118.86
89	7.943283×10^8	28183.83
90	1.000000×10^9	31622.78
91	1.258925×10^9	35481.34
92	1.584893×10^9	39810.72
93	1.995262×10^9	44668.36
94	2.511886×10^9	50118.72
95	3.162278×10^9	56234.13
96	3.981072×10^9	63095.73
97	5.011872×10^9	70794.58
98	6.309574×10^9	79432.82
99	7.943282×10^9	89125.09
100	1.000000×10^{10}	100000.0

Appendix E
Exponential growth and decay

The following table can be used to determine the voltages and currents present in C-R and L-R circuits. The *time* is specified in terms of the *time constant* of the circuit (equal to the product of C and R in the case of a C-R circuit or ratio of L to R in the case of an L-R circuit). The values given in the *growth* and *decay* columns are relative to the applied voltage or current (i.e. in the case of a capacitor charging in a C-R circuit, a value of 0.6321 is equivalent to 63.21% of the applied voltage).

Time	*Growth*	*Decay*
0.0	0.0000	1.0000
0.1	0.0951	0.9048
0.2	0.1812	0.8187
0.3	0.2591	0.7408
0.4	0.3296	0.6703
0.5	0.3934	0.6065
0.6	0.4511	0.5488
0.7	0.5034	0.4965
0.8	0.5506	0.4493
0.9	0.5934	0.4065
1.0	0.6321	0.3678
1.1	0.6671	0.3328
1.2	0.6988	0.3011
1.3	0.7274	0.2725
1.4	0.7534	0.2465
1.5	0.7768	0.2231
1.6	0.7981	0.2018
6.5	0.9984	0.0015

Time	*Growth*	*Decay*
1.7	0.8173	0.1826
1.8	0.8347	0.1652
1.9	0.8504	0.1495
2.0	0.8646	0.1353
2.4	0.9092	0.0907
2.6	0.9257	0.0742
2.8	0.9391	0.0608
3.0	0.9502	0.0497
3.2	0.9592	0.0407
3.4	0.9666	0.0333
3.6	0.9726	0.0273
3.8	0.9776	0.0223
4.0	0.9816	0.0183
4.5	0.9888	0.0111
5.0	0.9932	0.0067
5.5	0.9959	0.0040
6.0	0.9975	0.0024
8.5	0.9997	0.0002

Time	Growth	Decay
7.0	0.9990	0.0009
7.5	0.9994	0.0005
8.0	0.9996	0.0003
9.0	0.9998	0.0001
9.5	0.9999	0.0000
10.0	0.9999	0.0000

Example E.1

A C-R circuit comprising a 100nF capacitor and 10kΩ resistor is charged from a 5V supply. Determine the capacitor voltage after a time of 1.8ns.

The time constant of the arrangement must first be found.

time constant = C × R = 100nF × 10kΩ = $100 \times 10^{-9} \times 10 \times 10^{3}$ s $= 1 \times 10^{-9}$ s = 1 ns

The time at which we need to determine the capacitor voltage is 1.8ns from the start of charging and, since the time constant is 1ns, this is 1.8 times the time constant of the circuit. The required entry in the table can now be found by referring to the growth column at the point at which time is 1.8. This produces the value 0.8347 and hence the capacitor voltage at that instant will be 0.8347 × 5V = 4.17V.

Example E.2

A bleed resistor is to be wired in parallel with a reservoir capacitor of 47μF. If the capacitor is charged to 60V when the supply is connected, determine the value of the bleed resistor if the capacitor voltage is to fall to below 9V 10s after the supply is disconnected.

The proportion of the voltage remaining in the capacitor is given by 9V/60V or 0.15. Referring to the table where the decay value is 0.15 (or less) gives a time of 1.9. Thus the required interval of time is 1.9 times the time constant. Thus the time constant must be equal to (or less than) 10s/1.9 or 5.26s.

The value of bleed resistor may now be determined from:

$$\text{time constant} = C \times R$$

$$5.26\text{s} = 47\mu\text{F} \times R$$

thus

$$R = 5.26\text{s}/47\mu\text{F} = \frac{5.26}{47 \times 10^{-6}} = \frac{5.26}{47} \times 10^{6}\ \Omega$$

$$R = 0.11 \times 10^{6}\ \Omega$$

or

$$R = 110\text{k}\Omega$$

The nearest preferred value below this should be used to *ensure* that the capacitor discharges below the desired level within the time interval specified. A value of 100kΩ would thus be required.

Index